11th International Workshop NDT in Progress 2021

Prague, Czech Republic
4 – 7 October 2021

ISBN: 978-1-7138-4149-4

Printed from e-media with permission by:

Curran Associates, Inc.
57 Morehouse Lane
Red Hook, NY 12571

Some format issues inherent in the e-media version may also appear in this print version.

Printed with permission by Curran Associates, Inc. (2022)

For permission requests, please contact
Brno University of Technology & Czech Society for Non-Destructive Testing
at the address below.

Brno University of Technology
Techniká 2 CZ 616 69 Brno
Czech Republic

Phone: +420 541 147 664

www.cndt.cz

Additional copies of this publication are available from:

Curran Associates, Inc.
57 Morehouse Lane
Red Hook, NY 12571 USA
Phone: 845-758-0400
Fax: 845-758-2634
Email: curran@proceedings.com
Web: www.proceedings.com

TABLE OF CONTENTS

EUROPEAN NDT&CM2021 PRAGUE, CZECH REPUBLIC OCTOBER 4-7, 2021

BETTER CM & NDT – WARRANTY OF QUALITY, RELIABILITY AND SAFETY

FOUR NDT, CM AND SHM EVENTS IN ONE WEEK IN THE MIDDLE OF THE EUROPE

WE GO HYBRID!

Mechanical manipulator for use in hot cell Application of nonlinear method NWMS in irradiated concrete testing

Zbyněk Hlaváč*, Leoš Assmann, Jaroslava Zatloukalová; *) zbynek.hlavac@cvrez.cz

Research Centre Rez, Czech Republic

← Fig.1. Core boring in concrete slab (fabrication of samples)

Fig.2. → Encapsulation of concrete cylinders

Fig.3. Concrete samples treated by 400°C

Fig.4. Hydraulic press 300 kN in hermetic box with robotic arms

Fig.5. Manipulator Ester designed for work in radiation

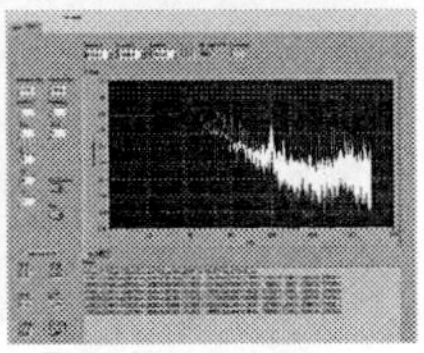

Fig.6. Software for non-destructive testing of concrete samples

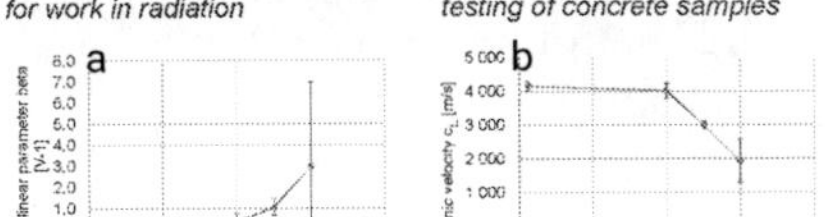

Fig.7. Nonlinear parameter beta a) and ultrasonic pulse velocity b) of concrete samples deteriorated by elevated temperatures t

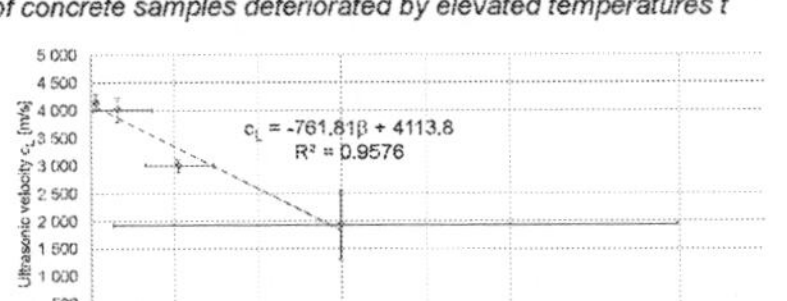

Fig.8. Ralation between nonlinear parameter beta and ultrasonic pulse velocity c_L

Acknowledgements: Presented work was executed under the financial support of VI20192022154 „Biological Shielding Concrete Testing after Reactor Core's Neutrons Irradiation", from the state budget through the Ministry of the Interior of the Czech Republic in Program Security research of the Czech Republic 2015-2022

Introduction

Energetics is one of the key sphere of industry. Nuclear energy supplies a large number of households or companies.

Safety problematic of nuclear power plant depends on its each critical component. Here the concrete of reactor's pressure vessel (RPV) cavity and methods of its testing are being discussed.

Methodology of non-destructive testing of RPV cavity made of concrete has been developed during last 6 years in Research centre Rez within grant project of the Ministry of Interior of the Czech Re-public. It used Nonlinear Wave Modulation Spectroscopy (NWMS) for identification of presence of cracks in concrete. These cracks occurred when the concrete was subjected to elevated temperature in the range of 105 to 600°C. Other type of degradation process was gamma irradiation or exposure to LOCA conditions, i.e. elevated pressure and temperature of surrounding atmosphere.

Current project solves damage of concrete's internal structure caused by neutrons and the verification of lastly developed methodology. Mechanical manipulator Ester was developed for non-destructive inspection of irradiated concrete samples. It was designed as part of robotic system used for manipulation with radioactive material in hot cells. Ester holds 3 piezoelectric transducers, two emitters and one receiver. These are attached to concrete surface using 3 springs and struts. All system can move up and down or rotate around tested sample what enables overall inspection of concrete including porosity, cavitation or heterogeneities.

Major irradiation experiment finishes in the year 2022, nevertheless presented paper informs about its progress and mid-term results.

Samples preparation and treatment

Initial fabrication started in 2019 when casting 4 great concrete slabs made of mixed Portland cement type CEM II, fine silica aggregates of diameter 2 ÷ 2.5 mm and tap water. Each slab cured in the same condition, i.e. in humid atmosphere in temperature range 22.5 ÷ 24°C. After 28 days three slabs were thermally deteriorated by temperatures 400, 500 and 600°C and one was left as reference. A year after, a large number of samples was drilled by core drill out of all mentioned slabs, Fig.1 and Fig.3. Reference samples made of reference slab were grinded and polished in order to obtain re-quested diameter 39.94 ÷ 40.00 mm. Then all these samples were dried carefully by temperature not exceeding 60°C with slow increment or decrement of temperatures, 20°C per day. After that half of the samples was encapsulated into aluminum tube and put into nuclear research reactor LVR-15 in Rez, Czech Republic, Fig.2. Second half of the samples was again left as reference in exicator, i.e. in dry conditions. Since that time, which is nearly a year, first half of the samples is being irradiated by neutron and gamma radiation, combined with dry and warm atmosphere (40 ÷ 50°C).

At the mean time other samples from all 4 slabs were let in laboratory conditions and tested non-destructively in chosen periods.

Manipulator design and fabrication

Mechanical manipulator Ester was developed for work in hot cells of Research Centre Rez, designed to resist high levels of gamma radiation.

Three ultrasonic transducers must be pressed to surface of concrete cylinder of diameter 30 ÷ 50 mm and height 40 ÷ 80 mm by certain force according to the needs of the operator. All transducers shall move around the sample, up and down. For this purpose, manipulator Ester was designed as cylindrical rotational cage with 3 radial sticks and one handle for its lifting.

Several difficulties had to be solves before manipulator fabrication. Firstly it wasn't possible to use lubrication in the manipulator. Irradiated particles could pollute lubricant as well as whole manipulator. Whole the instrument is to be operated by robotic arms without access of human operator, Fig.4. This was really as issue to be solved. And last but not least was the fact, that manipulator body should not exceed 5 kg weight because of limited capacity of the robotic arms.

Mobile upper part of the manipulator was made of lightweight aluminium alloy, stainless steel radial sticks were equipped by special springs determining contact force of transducers to concrete surface, Fig.5. Fixed lower part of the manipulator consists aluminium slab and inserted universal holder made of stainless steel. Total weight is more than 15 kg so that it cannot be moved by the robotic arm and makes stabile system for non-destructive testing.

First question after Ester's fabrication was whether it can attach the transducers onto concrete surface. This was proved in Brno, using digital oscilloscope Rigol DS1102E. Ultrasonic transducers were equipped by self vulcanizing tape and attached to the Rigol instrument. By hand driven signal was easily transmitted through the concrete sample into all 3 transducers. This confirmed Ester's ability of signal reading.

Second phase of Ester's testing was executed in Pilsen using ultrasonic testing device from National Instruments. Linear and nonlinear testing ultrasonic methods were applied there in order to prove Ester's readiness to operate in hot cell with Nonlinear Wave Modulation Spectroscopy NWMS. Some necessities still has to be added and fixed before manipulator's start in the hot lab. Even short presence in dusty radiation environment could disable Ester to return back to fabrication desk.

Nonlinear method NWMS and ultrasonic pulse method

Firstly, two acoustic signals (1 and 2) of different frequencies f_1 and f_2 are being emitted from waveform generator and amplified by high-power amplifiers. Both signals are modulated inside tested object, here concrete sample. Result signal is again pre-amplified and recorded by an oscilloscope. Level of intermodulation of both emitted signals can be described by nonlinearity parameter beta. The higher is β the bigger is concrete deterioration, e.g. higher number, length or width of cracks.

$$\beta = \frac{A_{12}}{A_1 \cdot A_2} \qquad (1)$$

β – parameter of nonlinearity of measured object,

A_{12} – measured amplitude of intermodulation of both emitted signals 1 and 2,

A_1 and A_2 – measured amplitudes of emitted signal 1 and 2.

Fig.7a shows mean values and standard deviations of nonlinear parameter β. For comparison, measured values of ultrasonic pulse velocity are in Fig.7b. Uniform conditions were applied for all samples.

The relationship between results of ultrasonic linear method and nonlinear method NWMS is noticeable in Fig.8. Both methods are sensitive to cracks. Moreover, ultrasonic method detects also voids or honeycombs and has narrow relation towards E-modulus of tested material. On the other hand, nonlinear method NWMS can recognize relatively very small cracking or even developments of micro-cracks in their initial state.

Destructive tests are planned for mid of the year 2022 when also the irradiated samples will be tested.

Conclusion

Goal of recent work is to develop and verificate NWMS method for non-destructive testing of highly irradiated concrete samples in hermetic enclosed box. One of the main outcomes is mechanical manipulator Ester designed for work in hot cell's radioactive atmosphere.

Here shown results declare ability of NWMS and ultrasonic pulse methods to detect non-homogeneities in concrete. Moreover, their combination can provide with an interesting idea about the type of defect or its dimension.

CENTRUM VÝZKUMU ŘEŽ S.R.O.

Hlavní 130, Řež, 250 68 Husinec, Czech Republic
T: +420 266 173 181, F: +420 266 172 398
E: cvrez@cvrez.cz, W: www.cvrez.cz

MINISTRY OF THE INTERIOR OF THE CZECH REPUBLIC

CVŘ Research Centre Řež

Mechanical manipulator for use in hot cell - application of nonlinear method NWMS in irradiated concrete testing

Zbyněk HLAVÁČ [1], Leoš ASSMANN [1], Jaroslava ZATLOUKALOVÁ [1]

[1] Centrum výzkumu Řež s.r.o., Husinec - Řež, Czech Republic; Phone: +420 723 946 561; e-mail: zbynek.hlavac@cvrez.cz, leos.assmann@cvrez.cz, jaroslava.zatloukalova@cvrez.cz

Abstract
Methodology of non-destructive testing of reactor pressure vessel cavity made of concrete has been developed during last 6 years in Research centre Rez within grant project of the Ministry of Interior of the Czech Republic (MVČR). It used Nonlinear Wave Modulation Spectroscopy (NWMS) for detection of cracks in thermally deteriorated concrete. Current project o MVČR solves damage of concrete's internal structure caused by neutrons and the verification of lastly developed methodology.
Mechanical manipulator Ester was developed for non-destructive inspection of irradiated concrete samples. It was designed as part of robotic system used for manipulation with radioactive material in hot cells.

Keywords: Nonlinear Wave Modulation Spectroscopy (NWMS), irradiated concrete, mechanical manipulator, hot cell, ultrasonic transducers, Ministry of Interior of the Czech Republic (MVČR), Light Water Reactor LVR-15

1. Introduction

Energetics is one of the key sphere of industry. Nuclear energy supplies a large number of households or companies.
Safety problematic of nuclear power plant depends on its each critical component. Here the concrete of reactor's pressure vessel (RPV) cavity and methods of its testing are being discussed.
Methodology of non-destructive testing of RPV cavity made of concrete has been developed during last 6 years in Research centre Rez within grant project of the Ministry of Interior of the Czech Republic. It used Nonlinear Wave Modulation Spectroscopy (NWMS) for identification of presence of cracks in concrete. These cracks occurred when the concrete was subjected to elevated temperature in the range of 105 to 600°C. Other type of degradation process was gamma irradiation or exposure to LOCA conditions, i.e. elevated pressure and temperature of surrounding atmosphere.
Current project solves damage of concrete's internal structure caused by neutrons and the verification of lastly developed methodology.
Mechanical manipulator Ester was developed for non-destructive inspection of irradiated concrete samples. It was designed as part of robotic system used for manipulation with radioactive material in hot cells. Ester holds 3 piezoelectric transducers, two emitters and one receiver. These are attached to concrete surface using 3 springs and struts. All system can move up and down or rotate around tested sample what enables overall inspection of concrete including porosity, cavitation or heterogeneities.
Major irradiation experiment finishes in the year 2022, nevertheless presented paper informs about its progress and mid-term results.

2. Experimental works

2.1 Samples preparation and treatment

Initial fabrication started in 2019 when casting 4 great concrete slabs made of mixed Portland cement type CEM II, fine silica aggregates of diameter 2 ÷ 2.5 mm and tap water. Each slab cured in the same condition, i.e. in humid atmosphere in temperature range 22.5 ÷ 24°C. After 28 days three slabs were thermally deteriorated by temperatures 400, 500 and 600°C and one was left as reference. A year after, a large number of samples was drilled by core drill out of all mentioned slabs, figures 1 and 2a. Reference samples made of reference slab were grinded and polished in order to obtain requested diameter 39.94 ÷ 40.00 mm. Then all these samples were dried carefully by temperature not exceeding 60°C with slow increment or decrement of temperatures, 20°C per day. After that half of the samples was encapsulated into aluminium tube and put into nuclear research reactor LVR-15 in Rez, Czech Republic, Fig. 2b. Second half of the samples was again left as reference in exicator, i.e. in dry conditions. Since that time, which is nearly a year, first half of the samples is being irradiated by neutron and gamma radiation, combined with dry and warm atmosphere (40 ÷ 50°C).
At the mean time other samples from all 4 slabs were let in laboratory conditions and tested non-destructively in chosen periods. Part of the testing will be introduced in chapter 3.

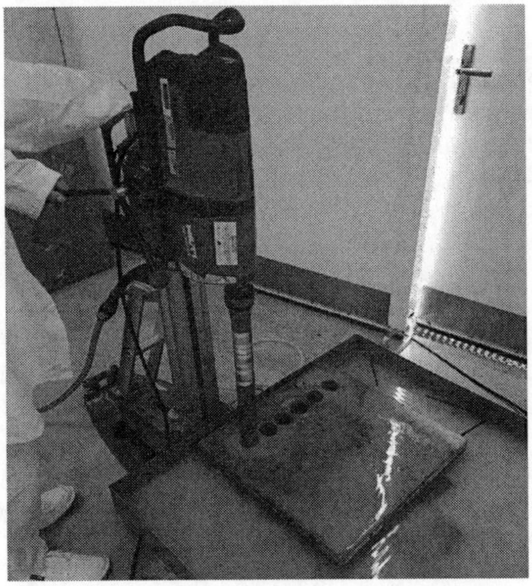

Figure 1. Core boring in concrete slab (fabrication of testing samples)

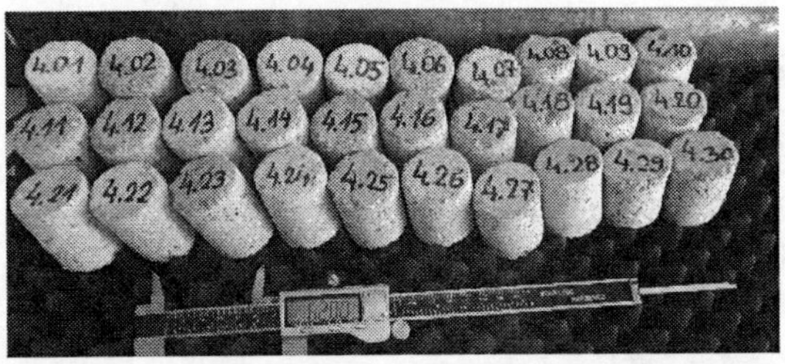

Figure 2a. Concrete samples treated by 400°C

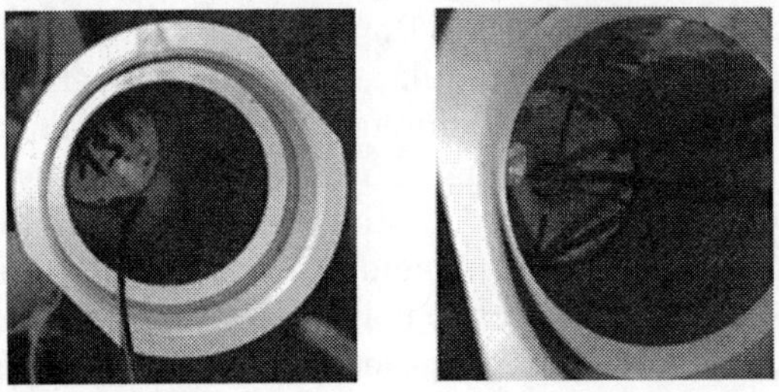

Figure 2b. Encapsulation of concrete cylinders before irradiation in research reactor LVR-15 in Rez

2.2 Manipulator design and fabrication

Mechanical manipulator Ester was developed for work in hot cells of Research Centre Rez, designed to resist high levels of gamma radiation.

2.2.1 Requested properties

Three ultrasonic transducers must be pressed to surface of concrete cylinder of diameter 30 ÷ 50 mm and height 40 ÷ 80 mm by certain force according to the needs of the operator. All transducers shall move around the sample, up and down. For this purpose, manipulator Ester was designed as cylindrical rotational cage with 3 radial sticks and one handle for its lifting.

2.2.2 Advanced design

Several difficulties had to be solves before manipulator fabrication. Firstly it wasn't possible to use lubrication in the manipulator. Irradiated particles could pollute lubricant as well as whole manipulator. Whole the instrument is to be operated by robotic arms without access of human operator, Fig 3. This was really as issue to be solved. And last but not least was the fact, that manipulator body should not exceed 5 kg weight because of limited capacity of the robotic arms.

2.2.3 Fabrication

Mobile upper part of the manipulator was made of lightweight aluminium alloy, stainless steel radial sticks were equipped by special springs determining contact force of transducers to concrete surface, Fig. 4. Fixed lower part of the manipulator consists aluminium slab and inserted universal holder made of stainless steel. Total weight is more than 15 kg so that it cannot be moved by the robotic arm and makes stabile system for non-destructive testing.

2.2.4 Manipulator testing

First question after Ester's fabrication was whether it can attach the transducers onto concrete surface. This was proved in Brno, using digital oscilloscope Rigol DS1102E. Ultrasonic transducers were equipped by self vulacanizing tape and attached to the Rigol instrument. By hand driven signal was easily transmitted through the concrete sample into all 3 transducers. This confirmed Ester's ability of signal reading.

Second phase of Ester's testing was executed in Pilsen using ultrasonic testing device from National Instruments. Linear and nonlinear testing ultrasonic methods were applied there in order to prove Ester's readiness to operate in hot cell with Nonlinear Wave Modulation Spectroscopy NWMS. Some necessities still has to be added and fixed before manipulator's start in the hot lab. Even short presence in dusty radiation environment could disable Ester to return back to fabrication desk.

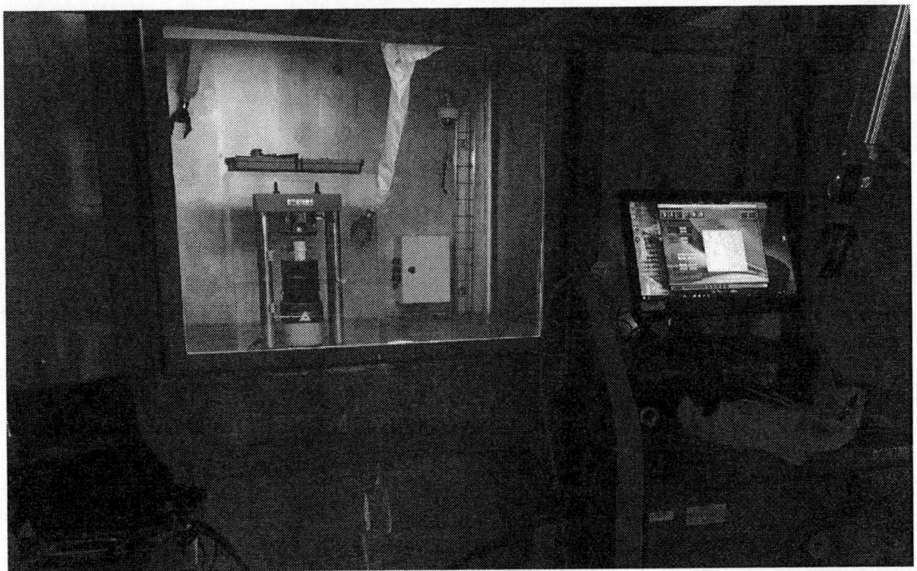

Figure 3. View into hermetic box 5b designed for radioactive concrete destructive and non-destructive testing

Figure 4. Prototype of mechanical manipulator Ester designed for work in radiation environment in hot cells

2.3 Nonlinear method NWMS and its application

Six years ago, nonlinear testing method NWMS was chosen for irradiated concrete testing. Czech as well as American studies showed capability of NWMS and other nonlinear ultrasonic methods to detect deteriorated concrete state. In comparison to other nonlinear acoustic methods, Nonlinear Wave Modulation Spectroscopy shows the degree of damage, which doubtless is an advantage.

2.3.1 NWMS principles

Firstly, two acoustic signals (1 and 2) of different frequencies f_1 and f_2 are being emitted from waveform generator and amplified by high-power amplifiers. Both signals are modulated inside tested object, here concrete sample. Result signal is again pre-amplified and recorded by an oscilloscope. Level of intermodulation of bot emitted signals can be described by nonlinearity parameter beta.

$$\beta = \frac{A_{12}}{A_1 \cdot A_2} \tag{1}$$

β – parameter of nonlinearity of measured object,
A_{12} – measured amplitude of intermodulation of both emitted signals 1 and 2,
A_1 and A_2 – measured amplitudes of emitted signal 1 and 2.
The higher is β the bigger is concrete deterioration, e.g. higher number, length or width of cracks. In highly damaged concrete, amplitude A_{12} cannot be measured at all. Poor acoustic conductivity can be caused by number of cracks in concrete.

2.3.2 Ultrasonic devices

Acoustic signals were generated by two arbitrary waveform generators (AWGs) PXI-5421 from National Instruments (NI) and amplified by two high-power amplifiers AR-150A100B from Amplifier Research.
Final signal was pre-amplified by preamplifier AS4koax from Dakel and received by 24-bit oscilloscope PXI-5922 (NI). All NI components were integrated in chasee NI PXIe-1078, figure 5.

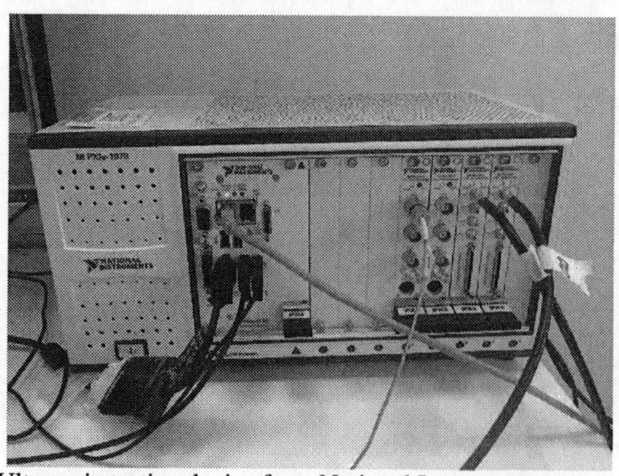

Figure 5. Ultrasonic testing device from National Instruments

2.3.3 Transducers

Two piezoelectric transducers MDK13-B were used for transmission of signal 1 and 2 into concrete sample. One transducer MDK13 -S was used for receiving the final signal from concrete surface. All three transducers were developed by Dakel, ZD Rpety, Czech Republic, as wide band gauges acoustic emission.

2.3.4 Testing software

Special software [1] was developed for work with ultrasonic device from National Instruments, figure 6. It uses programming platform Labview from National Instruments [2]. Its task is to set up frequencies f_1 and f_2 with amplitudes A_1' and A_2', generate both signals, read and store their results, A_1, A_2 and A_{12} at frequency $f_{12} = f_1 + f_2$ and calculate parameter beta using formula (1).

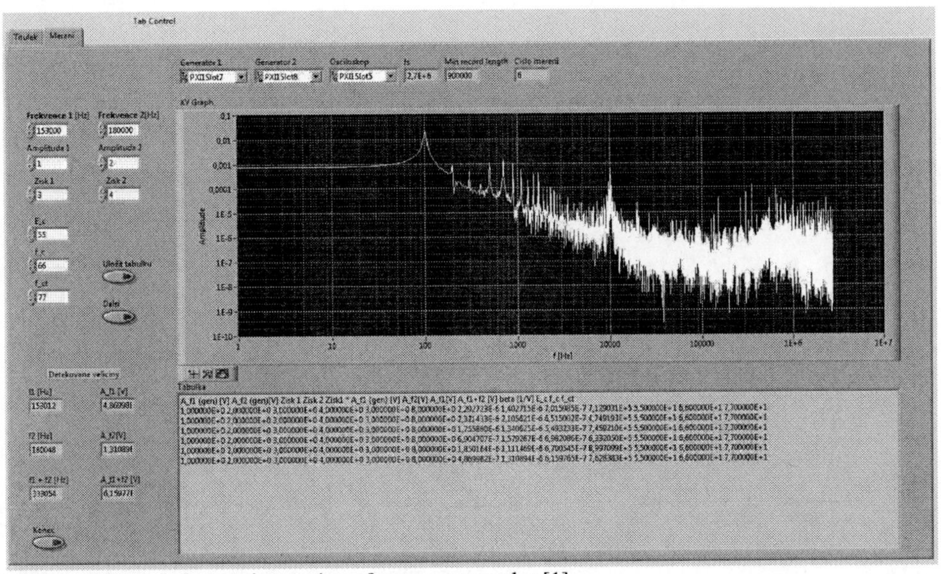

Figure 6. Software for non-destructive testing of concrete samples [1]

2.4 Experimental results

Four groups of concrete samples were tested using mechanical manipulator Ester: reference samples (24°C treatment), 400°C, 500°C and 600°C. Fabrication, curing conditions and thermal treatment are described in chapter 2.1. Uniform conditions were applied for all group of samples. Input signals were set up as f_1 = 153 kHz, A_1 = 2 V and f_2 = 180 kHz, A_2 = 2 V as well.
Figure 7 shows mean values and standard deviations of nonlinear parameter β.
For comparison, measured values of ultrasonic pulse velocity is in Figure 8. Ultrasonic pulse method and its application in thermally damaged concretes was described for example in [3].
Destructive tests are planned for mid of the year 2022 when also the irradiated samples will be tested.

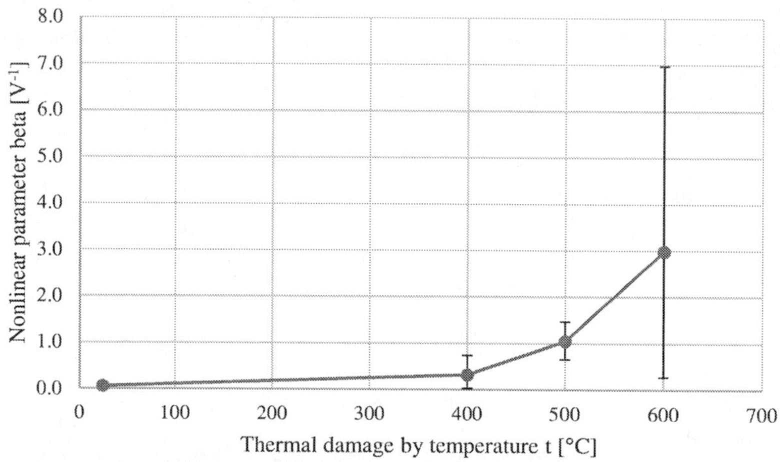

Figure 7. Nonlinear parameter β of concrete samples treated by 24, 400, 500 and 600°C

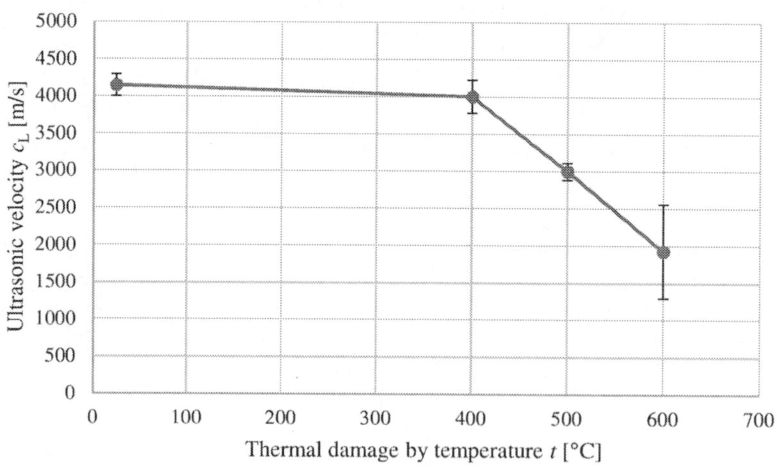

Figure 8. Ultrasonic pulse velocity in concrete samples treated by 24, 400, 500 and 600°C

The relationship between results of ultrasonic linear method and nonlinear method NWMS is noticeable in Figures 9. Both methods are sensitive to cracks. Moreover, ultrasonic method detects also voids or honeycombs and has narrow relation towards E-modulus of tested material. On the other hand, nonlinear method NWMS can recognize relatively very small cracking or even developments of microcracks in their initial state.

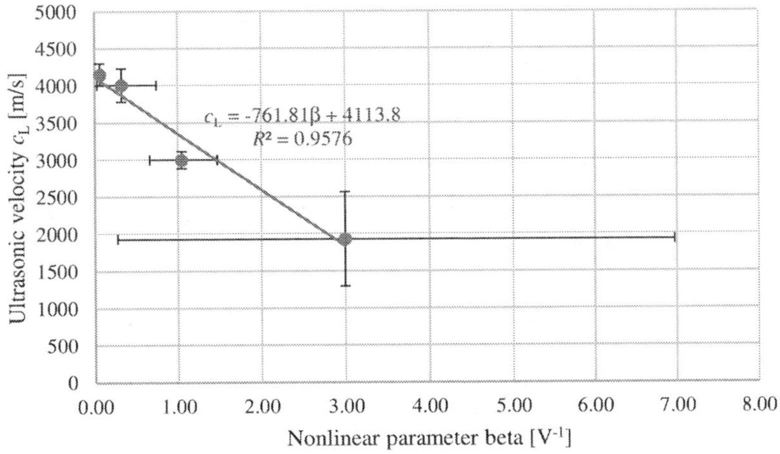

Figure 9. Prediction of ultrasonic velocity c_L on the base of nonlinear parameter beta

3. Conclusions and remarks

Paper introduced goals of solved project, its progress and interim results. Main goal is development and verification of method NWMS for non-destructive testing of highly irradiated concrete samples in hermetic enclosed box. One of the main outcomes of the project is mechanical manipulator Ester designed for work in radioactive atmosphere in a hot lab.
Here shown results declare ability of NWMS and ultrasonic pulse methods to detect inhomogeneities in concrete. Moreover, their combination can provide with an interesting idea about the type of defect or its dimension.

Acknowledgement
This paper was prepared under the project VI20192022154 'Biological Shielding Concrete Testing after Reactor Core's Neutrons Irradiation' with financial support from the state budget through the Ministry of the Interior of the Czech Republic in Program Security research of the Czech Republic.

References

1. J Jansa and J Patera, 'Uživatelský manuál k programu pro nedestruktivní analýzu ozářených betonových vzorků', Technical report, Centrum výzkumu Řež s.r.o., February 2021.

2 National Instruments – NI LabVIEW user manual. http://www.ni.com/pdf/manuals/320999e.pdf

3. Z. Hlaváč, J Patera and P Halodová, 'Jak a proč se v Řeži zkouší beton?', Jaderná energie 2020, 66 (SI) , 106-111.

Fast inspection through arbitrary shape interfaces with Total Focusing Phased Array

Guillermo COSARINSKY [1,3,4], Jorge F. CRUZA [2], Ricardo GONZÁLEZ [2], Jorge CAMACHO [1]

[1] Ultrasound Systems and Technology Group (GSTU), Institute for Physical and Information Technologies (ITEFI), Spanish National Research Council (CSIC), c/Serrano 144, 28006 Madrid, Spain; g.cosarinsky@csic.es; j.camacho@csic.es

[2] DASEL SL, Avda. del Cañal 44 Nave 3, 28500 Madrid, Spain; cruza@daselsistemas.com, ricardo@daselsistemas.com

[3] Non Destructive Testing Department, National Atomic Energy Commission (CNEA), Av. Gral. Paz 1499, Buenos Aires B1650, Argentina

[4] Electronics Department, Escuela Politécnica, Universidad de Alcalá de Henares, Ctra.Madrid-Barcelona, Km. 33,600, 28805 Madrid, Spain

Abstract
Using the Phased Array (PA) technique for inspecting large components in water immersion can be particularly challenging. Little changes of array position or component shape require to recalculate the focal laws to maintain the programmed scan, because of beam refraction. These focal laws should be calculated and programmed on-the-fly, which is not usually implemented in PA equipment and, when done, it limits the inspection speed.
Total Focusing Phased Array (TFPA) is an ultrasound imaging method similar to traditional Phased Array (PA) but achieving dynamic depth focusing both in emission and reception. The underlying idea is to develop an imaging method that complies with PA inspection standards (definition of A-Scan lines, scan angles, etc.) but with an equivalent image quality to that of Total Focusing Method (TFM).
TFPA has an additional advantage over PA. The acquisition scheme is Full Matrix Capture (FMC), and hence, there is no need to compute emission focal laws in real time. In this work, we acquire a sequence of FMC captures of an immersed component, while mechanically moving the array at high speed. Images are then computed in post-processing using a Graphical Processing Unit (GPU).
Keywords: Ultrasound imaging, Phased Array (PA), Total Focusing Method (TFM), Inspection, Full Matrix Capture (FMC), Graphical Processing Unit (GPU)

1. Introduction

Ultrasound inspection of large component is usually performed using mechanical automated (or semi-automated) devices, such as linear manipulators, bridge designs or robotic manipulators [1]. For components where direct contact testing is not possible, immersion testing is one of the alternatives. Local immersion or water filled flexible wedges are frequently used. In this cases refraction at test piece surface must be accounted for in Time Of Flight (TOF) computations, and surface estimation must be done to compute ray paths.

As the array position and component shape might change along the scanning displacement, for PA inspection it is required to recalculate the focal laws to maintain the programmed scan. These focal laws should be calculated and programmed on-the-fly, which is not usually implemented in PA equipment and, when done, it limits the inspection speed.

An alternative to PA is FMC acquisition. Unlike PA, where a sequence of focal laws is used to manipulate the ultrasound beam, in FMC acquisition is done by firing sequentially each array element and recording the A-scans received by all elements [2]. The data collected is an array with size (NE^2 x NS), being NE the number of array elements, and NS the number of samples in each A-scan. This array is used to compute images with TFM. This images are composed of a rectangular (NZ x NX) pixel grid spanning a Region Of Interest (ROI) (Figure 1a). TOF from each array element to each image pixel must be computed in order to sum coherently the A-scan amplitudes. TFM achieves focusing in every pixel, both in emission and reception. It is considered the "gold standard" in ultrasound imaging.

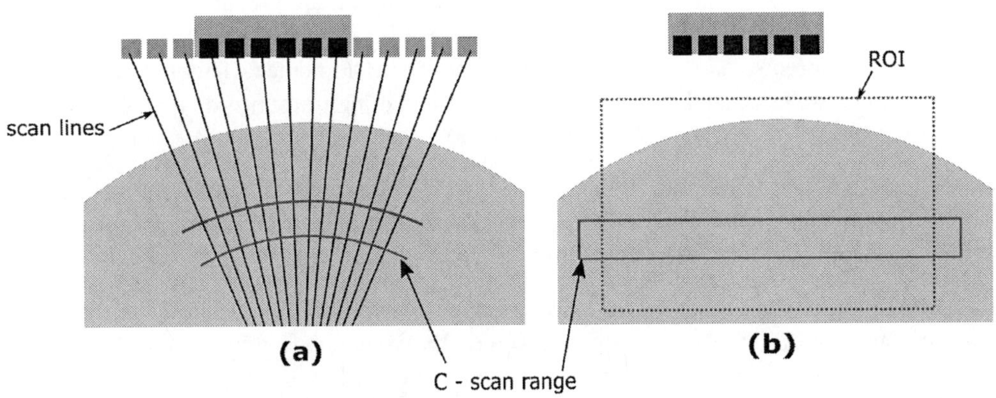

Figure 1. Image geometry. a) TFPA, image is computed on scan lines like those of PA; b) TFM, image is computed on a rectangular pixel grid

2. The Total Focusing Phased (TFPA) Array method

In immersion testing, ultrasound refraction at the test piece surface has to be accounted for the computation of TOF. Fermat's principle is used to find the entry point of each ray. For this calculation an iterative search is needed. Thus, TFM is a computationally intensive task.

The Virtual Array (VA) method [3] is a technique to simplify TOF calculations in PA inspections of immersed components.. In this method, for each of the NL scan lines, a virtual source is computed for each array element. This virtual source behaves like a source in the second propagation, hence, the VA dynamic depth focusing in reception can be done in real time. The details of the procedure to compute the coordinates of each virtual source are presented in [3].

To compute the virtual array coordinates for a scan line, the exact TOF must be evaluated only for to foci in the line, instead of the NR samples on the line. This reduces the number of iterative searches by a factor of NR/2, reducing significantly the computational load.

TFPA, as presented in [4] is somehow like a mixture of PA and TFM. The acquisition used is FMC. But unlike TFM, the image is computes on a set of scan lines as those of PA (figure 1b). Then, the VA method can be used just like in PA to compute TOF for each point on each scan line. Thus, the image is focused both in emission and reception as in TFM.

In PA, little changes of array position or component shape require to recalculate the focal laws to maintain the programmed scan, because of beam refraction. These focal laws should be calculated and programmed on-the-fly, which is not usually implemented in PA equipment and, when done, it limits the inspection speed. In TFPA, the test piece surface can be estimated from the FMC data in post-processing, so more precise and robust algorithms can be implemented. This is not the case in PA, where surface must be estimated on-the-fly using simple enough algorithms that don't limit the inspection speed.

As mentioned before TFPA computes images along scan lines as those of PA. Both linear and sectorial scans can be implemented, which gives the method the possibility to comply with PA inspection standards (definition of A-Scan lines, scan angles, etc.).

A special feature of TFPA, is that it can produce images with more lines than those we can measure with the same array using PA. As illustrated in figure 1a, we can define lines along "beams" coming from hypothetical elements (green elements in figure 1a) outside the array, as if it had more than N elements. This can be achieved because the acquisition mode is FMC, where the unfocused emission of individual elements can reach points outside the "shadow" of the array.

3. Experimental

A brass test piece (figure 2a) was scanned in an immersion tank using a linear translation device at a 150 mm/s speed. The piece is a tube with 3 cylindrical sections of different radii. This changing cross section provides the scenario to test the performance of TFPA when shape changes and misalignment are present. The test piece was laid manually in the tank with its axis approximately in the scanning direction, so there must be some amount of misalignment.

The test piece has artificial defects simulating corrosion, as show in figure 2b.

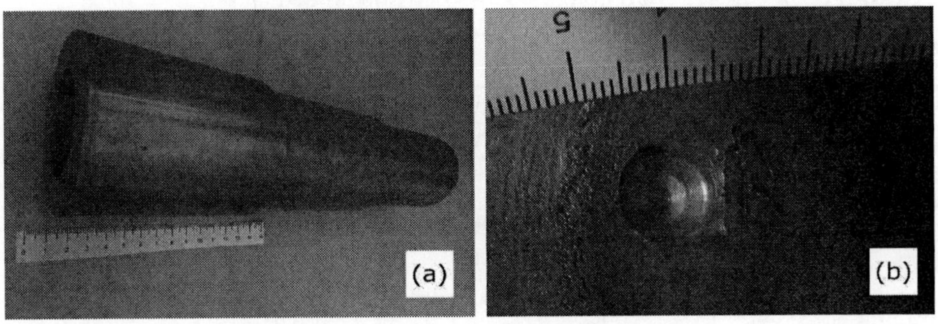

Figure 2. a) test piece; b) artificial defect detail

An acquisition of 270 FMC captures was done along a displacement of 270 mm. The equipment used was a 5 MHz, 32 elements, 0.6 mm pitch array and the Sitau II ultrasound system (DASEL SL, Madrid, Spain), which has a Gigabit Ethernet connection to the PC. The full acquisition was transferred to PC memory in about 5 s. Scanning parameters are shown in table 1.

The TFM and TFPA images where computed in a PC with AMD Ryzen 9 3900X 3.79 GHz 12-Core processor and a Graphical Processing Unit (GPU) GeForce RTX 3080. In house software developed by the authors was used.

4. Results

The post-processing speeds achieved are shown in table 1. To compute 270 TFM images the time taken was 11 s, and 4 s was the time required for the 270 TFPA images, about 3 times faster than TFM.

The total time to inspect the test piece, from displacement start to whole test piece TFPA image, was about 12 s.

Table 1. Scanning parameters and image post-processing speed

Scanning parameters			
Array	Scan speed	Number of FMC/mm	Transfer rate to PC memory
5 MHz 32 elements 0.6 mm pitch	150 mm/s	1	≈ 50 Mb/s

Images		
	Image size	Post-processing image rate
TFM	NZ = 230, NX = 300	22 fps
TFPA	NR = 400, NL = 100	54 fps

Results shown in subsequent figures use a coordinate system where x is the active array direction, y the passive direction, and z the array normal direction. Figure 3 shows a D-scan of the test piece, where defects are clearly distinguished. Interface and back-wall echoes are observed too. The interface points detected and used for TOF calculation are show in figure 4, where axial misalignment is observed in the top view.

Figure 5a and figure 5b show respectively a TFPA and a TFM B-scan corresponding to the section indicated by the red vertical line in figure 3. There is no significant difference between the two images.

Finally, figure 6 shows a C-scan corresponding to the region marked by a red rectangle in figure 3.

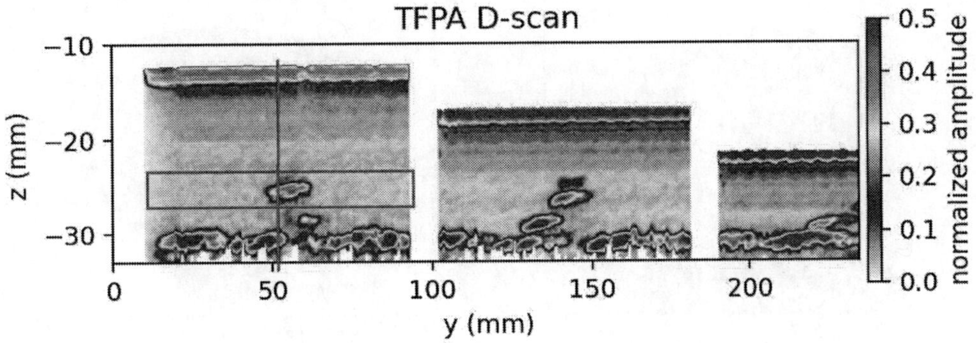

Figure 3. TFPA D-scan. The vertical red line indicates de section of the B-scan shown in figure 5. The red rectangle indicates the region of the C-scan shown in figure 6.

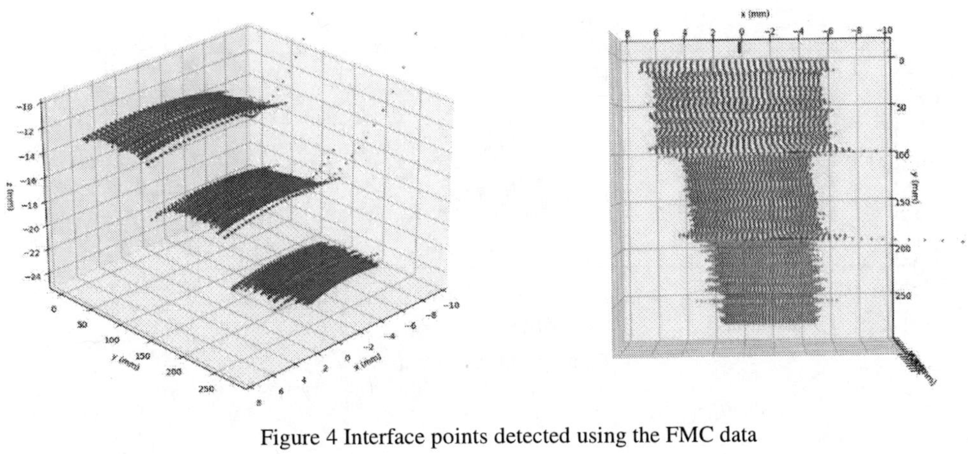

Figure 4 Interface points detected using the FMC data

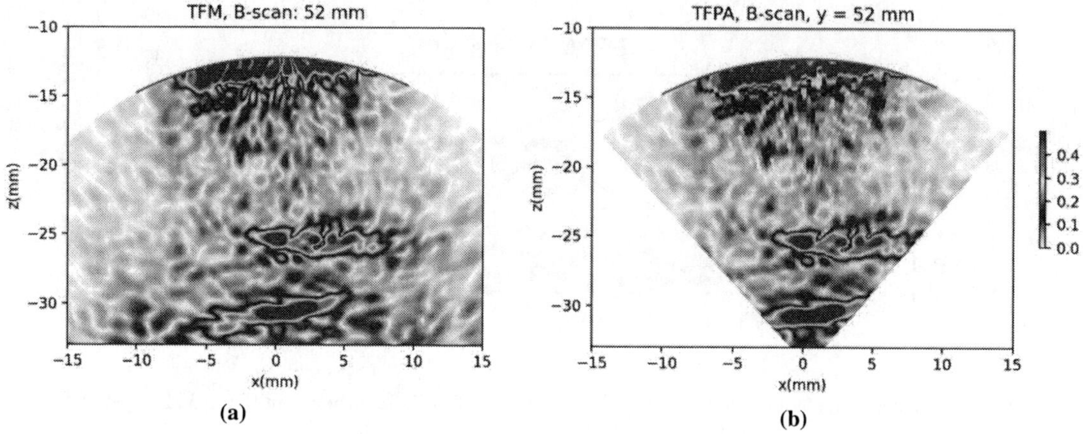

Figure 5. B-scans. a) TFM; b) TFPA

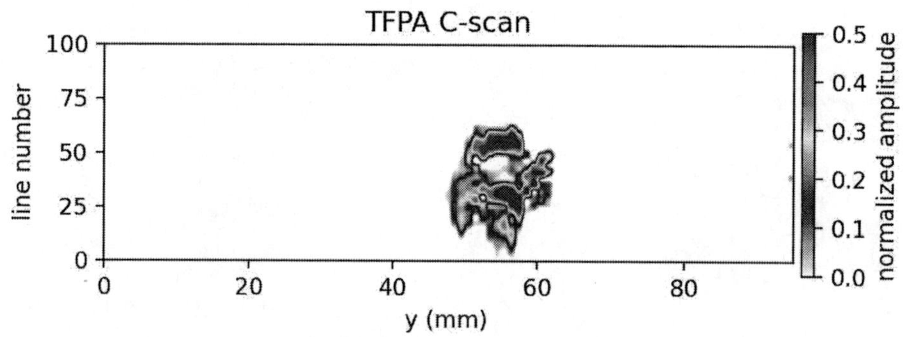

Figure 6. TFPA C-scan

5. Conclusions

A methodology for fast ultrasonic array inspection of large pieces was proposed. The method consists in a mechanical scan at 150 mm/s using FMC acquisition. TFPA images were computed in post-processing at 54 fps. The total time to inspect a 270 mm test piece and compute totally focused images at a 1 mm step was about 12 s.

Although the method does imaging in a post-processing stage, it could be adapted to work in real-time if a lower scanning speed is used. In both cases, the use of FMC acquisition allows for refinement of the images by adjusting imaging parameters.

Acknowledgements

This work is supported by the project RTI2018-099118-A-I00 founded by MCIU/AEI/FEDER, UE.and the fellowships PTQ-17-09089 and PRE2019-088602 of the Spanish Ministry of Science and Innovation.

References

1. C. Mineo et al., “Fast ultrasonic phased array inspection of complex geometries delivered through robotic manipulators and high speed data acquisition instrumentation,” IEEE Int. Ultrason. Symp. IUS, vol. 2016-Novem, pp. 16–19, 2016, doi: 10.1109/ULTSYM.2016.7728746.

2. C. Holmes, B. W. Drinkwater, and P. D. Wilcox, “Post-processing of the full matrix of ultrasonic transmit-receive array data for non-destructive evaluation,” NDT E Int., vol. 38, no. 8, pp. 701–711, 2005, doi: 10.1016/j.ndteint.2005.04.002.

3. J. F. Cruza, J. Camacho, L. Serrano-Iribarnegaray, and C. Fritsch, “New method for real-time dynamic focusing through interfaces,” IEEE Trans. Ultrason. Ferroelectr. Freq. Control, vol. 60, no. 4, pp. 739–751, 2013, doi: 10.1109/TUFFC.2013.2622.

4. J. F. Cruza, J. Camacho and J. M. Moreno, "Total Focusing Phased Array," *2016 IEEE International Ultrasonics Symposium (IUS)*, 2016, pp. 1-4, doi: 10.1109/ULTSYM.2016.7728610.

Approaches Combining Multiple Paths to Establish the Probability of Detection of a Guided Wave-based Structural Health Monitoring System

Ahmed B. M. BAYOUMI [1], Inka MUELLER [1], Thomas VOGT [2], Peter KRAEMER [3]
[1] Hochschule Bochum, Mechatronics and Mechanical Engineering Department, 44801 Bochum, Germany
Phone: +49 1766 8607217; e-mail: ahmed.bayoumi@hs-bochum.de, inka.mueller@hs-bochum.de
[2] Guided Ultrasonics Ltd, Brentford, TW8 8HQ, United Kingdom; e-mail: thomas.vogt@guided-ultrasonics.com
[3] University of Siegen, Mechanical Engineering Department, 57076 Siegen, Germany.; e-mail: peter.kraemer@uni-siegen.de

Abstract
This paper presents three different approaches combining Damage Indices (DIs) of multiple paths to generate a Probability of Detection (POD) curve and compares them with a single path-based data analysis technique. The three approaches are: 1) maximum value, 2) mean of maximum six paths, and 3) pattern analysis of DIs from all paths. A correlation coefficient-based DI is calculated for all paths. The approaches are tested using a data set provided by Open Guided Waves (OGW) platform. The structure used is a composite structure. The guided wave-based SHM system consists of 12 transducers. The damage used is a pseudo fault of an added mass with damping tacky tape. The results show a difference between the $a_{90|95}$ values of the three approaches to combine DIs of multiple paths. The combination of multiple paths can quantify the detection capability for a whole SHM system.
Keywords: structural health monitoring (SHM), guided waves, probability of detection (POD), threshold, damage index (DI), and probability of false alarm (PFA), openguidedwaves.de.

1. Introduction

SHM systems make a promising technique for monitoring and evaluating the integrity of many industrial structures like pipelines, aircraft, nuclear powerplant, and civil structures [1]. Composite structures used in aircraft must continue in flight-safe conditions even with stronger impacts [2]. These substantial impacts cause visible and invisible damage. The aircraft industry could benefit significantly from using SHM systems for damage detection for many reasons: increased safety due to continuous monitoring [3], accessibility to areas difficult to reach for inspection as the sensors are already embedded in the structure, reduced maintenance as it will be scheduled only when required [4], and improved the design of aircraft since the critical area will automatically be monitored.

Many guided wave-based SHM systems compare the current measurements with a baseline response that is considered pristine or damage-free by calculating DI to determine the current state of the structure [5].

The reliability assessment of SHM systems provides information regarding damage detection capability at certain confidence bounds. The methods used for assessing the reliability of an SHM system are originated from the guidelines for conventional non-destructive inspection techniques. The POD curve, as a reliability assessment technique, is a cumulative distribution curve generated from a large number of measurements. These measurements are recorded at different damage sizes. A generated POD curve provides information regarding the smallest damage size that can be reliably detected or the biggest damage size that may be missed [6]. The probability of false alarm (PFA) and threshold determination play a significant role in successfully generating a POD curve.

There are some differences between NDT and SHM systems that change the interpretation of POD curves. Guided wave testing as an NDT method is based on handheld movable sensor arrays, whereas guided wave testing as an SHM method is based on permanently installed transducers with fixed sensor positions. Despite the help of artificial intelligence, data analysis in NDT requires well-trained NDT personnel, while SHM system developers have the firm intention to enable automated inspection and data evaluation.

Using guided wave-based SHM systems on a composite structure is quite challenging. Measurements are recorded from a network of transducers, where each recorded measurement is along a path between a pair of transducers. Path-based data analysis is one approach used to generate POD curves, with each POD curve representing one path. Especially if the systems consist of many transducers, the path-based approach reaches its limits.

This paper presents several approaches to generate a POD curve and to compare it with a single path-based data analysis. These approaches are mainly focused on statistically combining all the information gathered from transducer pairs. The probability of detection section presents a brief review of POD curve generation. Furthermore, a case study using data from an open guided wave platform [7], its results and discussion are presented. Finally, the conclusion is presented.

2. Probability of detection

SHM systems can be briefly described at four levels: detection, localization, characterization, and prognostics [8]. Although all levels are of interest for an SHM system, this work focuses only on detection (Level 1). The POD curve is used to estimate the damage detection capability of an SHM system. The basic POD concept is documented in detail in [9], [10], [11] and [12]. Different methods are applied to assess the detectability of damage: the hit/miss method and flaw size method vs. response ($\hat{a}$ vs. a). The hit/miss method produces a binary statement or qualitative information about the presence or absence of damage. In contrast, the ($\hat{a}$ vs. a) method provides some quantitative measures.

$a_{90|95}$ is the damage size that can be detected with a probability of detection of 90% with a 95% Wald confidence interval. $a_{90|95}$ is estimated using the POD curve [13]. The typical procedure, considered to estimate $a_{90|95}$ using guided wave-based SHM system, consists of data processing, calculation of DI, threshold determination based on undamaged training data, regression analysis, and POD curve generation.

Once the data has been recorded and filtered, the calculation of DI takes place. It is a comparison between the baseline measurement and other recorded measurements. Calculation of DI provides information regarding each recorded measurement, whether from a pristine or a damaged structure. Several types of damage indices can be calculated: DIs in the time domain, in the time-frequency domain, and in the frequency domain [14].

The following step is the threshold ($\hat{a}_{threshold}$) determination. For known damage and undamaged data, the threshold value is the intersection between its distribution curves. Otherwise, the threshold determination is based on undamaged data only. Here it is assumed to have normally distributed DIs of the undamaged data:

$$\hat{a}_{threshold} = \mu + \sigma \cdot \phi^{-1}(1 - PFA), \tag{1}$$

Where ϕ is Gaussian cumulative distribution function of these reference data, μ and σ are the mean value and standard deviation of DIs of pristine measurements, and PFA is the probability of false alarm. Below the threshold, the state of the structure is considered undamaged. Once the value exceeds the threshold level, the structure is considered damaged. The higher the threshold, the bigger the smallest damage that can be detected. The lower the threshold, the higher the PFA, as shown in figure 1.

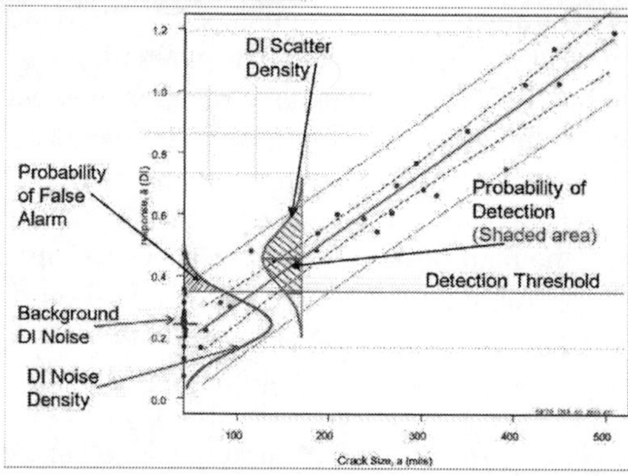

Figure 1. Influence of PFA on estimation of the threshold. As it decreases, the threshold detection increase [15]

After calculation of the DIs and determination of threshold, the functional relationship, preferably linear relationship, between ($\hat{a}$ vs. a) is established by means of Regression analysis. The green line, in figure 1, represents linear relationship between ($\hat{a}$ vs. a). The regression equation for a line of best fit to a given data is:

$$y = mx + \beta_1 + \beta_2, \tag{2}$$

Where $y = f(\hat{a})$ and $x = f(a)$. m, and β_1 are the slope and intercept of the linear relationship. β_2 is the standard deviation of the scattering around the regression line. It is caused by parameters other than a having an influence on $\hat{a}$, like temperature changes and different operational conditions. Then, the Delta method, a statistical technique for deriving a variance of the model parameters, is applied to link the signal to POD. It is a transition from $\hat{a}$ vs. a data to POD vs. a data [16]. Finally, a POD curve is generated. Regression analysis is used to estimate m, $\beta_1$and β_2 to calculate σ_{POD} and μ_{POD}:

$$\sigma_{POD} = \frac{\beta_2}{m}, \mu_{POD} = \frac{\hat{a}_{threshold} - \beta_1}{m}, \tag{3}$$

Hence, the POD curve can be generated using:

$$POD = [\frac{x - \mu_{POD}}{\sigma_{POD}}], \tag{4}$$

This POD curve is generated with a 50% confidence bound. The 95% confidence bound is achieved by computing the covariance matrix for the POD parameters (σ and μ).

Finally, $a_{90|95}$ can be estimated:

$$a_{90|95} = \mu_{a|POD=90\%} + z(\alpha)\sigma_{a|POD=90\%}, \tag{5}$$

For this case, $z(\alpha)$ is 1.645 and represents the confidence bound value of 95% for on tailed standard normal distribution.

3. Case study: OGW platform

The data set used for the POD calculation is available at Open Guided Waves (OGW) online platform under 'guided wave basic measurement data for the plate with a stringer' [7].

3.1 Structure layout

A square plate of length 500mm with an omega stringer, both made of Carbon Fiber Reinforced Polymer (CFRP) materials, are used as a structure. 12 piezoceramic transducers (T_1 - T_{12}) were bonded to the plate as shown in figure 3(a).

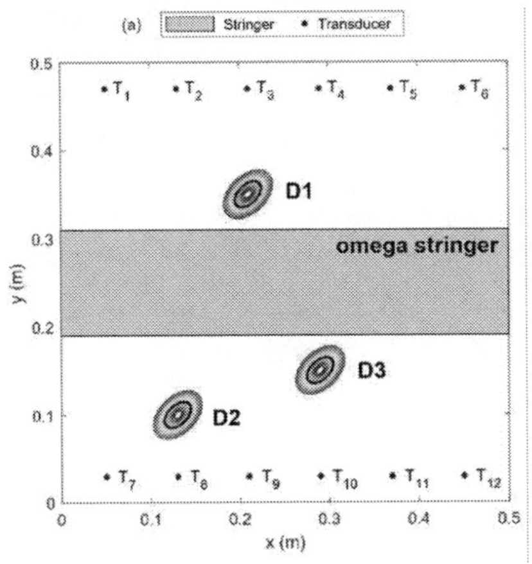

(a)

(b)

Figure 3. (a) Specimen geometry. Visualization of the 3 damage positions (D1-3) with respect to transducer location (T_1- T_{12}), and omega stringer over the square plate (b) Reference damage geometry [7]

3.2 Reference damage

Thirteen steel metallic discs with different sizes, as shown in figure 3(b) and table I, were attached to the CFRP structure by tacky tape. Adding mass to the structure causes a stiffness asymmetry by changing the structure's geometry and an additional damping depending on its location. Despite its simplified shape, the interaction with guided waves leads to a decrease in amplitude, a change in time of flight, and a mode conversion. The reference damage was positioned at D1, D2, and D3 individually.

Table I. Reference damage size [7]

Number	Damage size [mm^2]
1	49.48
2	111.33
3	197.92
4	309.25
5	445.32
6	606.13
7	671
8	791.68
9	1001.97
10	1237
11	1496.77
12	1781.28
13	2090.53

3.3 Experimental setup

The structure layout was placed in a climate chamber to maintain humidity and temperature at 50% and 23°C. Two pt-100 sensors were attached to the surface of the plate on the top and bottom to detect any thermal gradients.

For guided waves actuation and acquisition, piezoceramic transducers were used to actuate a tone burst signal and record the measurement. The voltage amplitude was defined as $\widehat{U} =$ 100V. The centre frequencies (f_c) were chosen to be $f_c = 40 - 260$kHz with an increasing step of 20kHz. The actuation waveform is a tone burst sine signal of five cycles Hann window, as shown in figure 4.

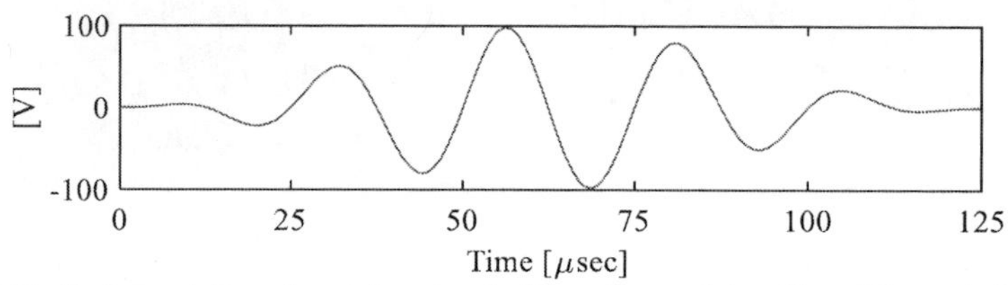

Figure 4. Excited signal. Tone burst sine signal of 5 cycles modulated by a Hann window at 40kHz

3.4 Recorded measurement

The total data set of (OGW) platform is 85 measurements recorded for each frequency at each damage position (D1 − 3). These measurements consist of 20 pristine measurements and 65 damaged ones (5 measurements for each damage size). Each measurement contains data recorded from 12 transducers. These results record 66 paths. Each path refers to an actuator-sensor pair.

Figure 5(a) shows the recorded measurement from the pristine structure and the damaged structure. As can be seen, the amplitude of the damaged signal, the red line, is slightly lower than the amplitude of the pristine.

The differential signal, as shown in figure 5(b), plays a very important role in clearly differentiate between the damaged signal and the pristine one. It is a subtraction of baseline measurement from the recorded measurements. The amplitude of the differential signal of the damaged measurement is more than twice the amplitude of the pristine one.

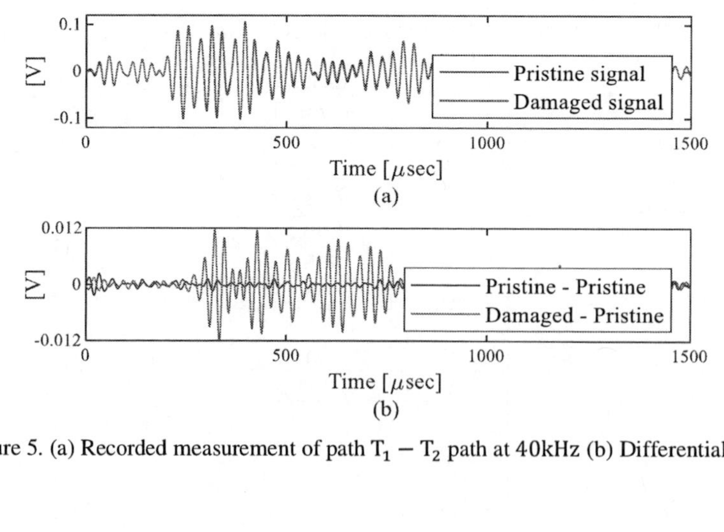

Figure 5. (a) Recorded measurement of path $T_1 - T_2$ path at 40kHz (b) Differential signal

4. Results and Discussion

The results shown in this section are related to $f_c = 40\text{kHz}$ and reference damage located at D1 only, however, Table III provides information regarding $a_{90|95}$ values located at D1, D2, and D3.

The post-processing analysis of the recorded measurements took place over a part of the signal after being filtered using bandpass filter within $[25 - 75\text{kHz}]$ interval. The first $125\mu\text{sec}$ of the recorded measurements is affected by cross-talk between the channels during exciting the signal. The duration of the excited signal is highly dependent on the frequency and its number of cycles. Moreover, by having a closer look over the differential signal, shown in figure 5(b), the dominant amplitude difference in the differential signal was within $750\mu\text{sec}$. Based on that, the post-processing analysis is based on a time window of $[125 - 750\mu\text{sec}]$, as shown in figure 6.

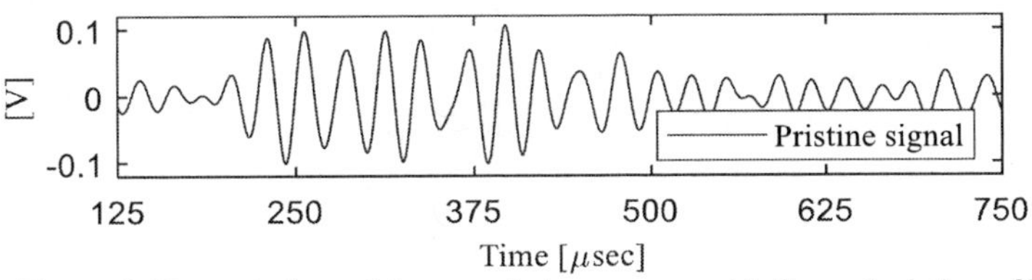

Figure 6. Time window of the recorded measurement before calculation of DI

Thanks to multiple paths, the calculation of a correlation coefficient-based DI [17] between a pristine measurement and a damaged one results in 66 DI values, one for each path.

$$\text{DI} = 1 - \frac{\sum_{i=1}^{N} S_{\text{H},i} S_{\text{D},i} - \sum_{i=1}^{N} S_{\text{H},i} \sum_{i=1}^{N} S_{\text{D},i}}{\sqrt{\sum_{i=1}^{N} S^2{}_{\text{H},i} - (\sum_{i=1}^{N} S_{\text{H},i})^2} \cdot \sqrt{\sum_{i=1}^{N} S^2{}_{\text{D},i} - (\sum_{i=1}^{N} S_{\text{D},i})^2}}, \tag{6}$$

Where S_H and S_D are the damaged signal and the pristine one respectively. i and N are the acquisition points that represent the start and the end of the time window. It was found that some of these paths are insensitive to damage located in D1. In these cases, using path-based analysis yields unrealistic POD curves, as the curves are shifted to values of a, which is larger than the plate itself. Furthermore, poor measurements will severely affect POD curves. Combining multiple paths is therefore proposed as a statistical approach with the intention to decrease the impact of potentially poor measurements, for example low sensitivity experienced in some paths.

Three different combining approaches are proposed and are compared to the single path-based analysis data analysis. The choice of one path has been determined carefully based on the maximum value of DI analysis in all measurements. More than 50% of DI values obtained of path $[\text{T}_3 - \text{T}_8]$ for damage located at D1. The three proposed approaches are:

- The maximum value analysis (M1) is the maximum value of DI values in each measurement. That means that the values considered are not necessarily from the same path in each measurement. Figure 8(a) represents the DI values of M1 considered for all measurements.
- The mean value of maximum six paths (M2): six paths considered roughly the highest 10% DI values in each measurement. The resultant DI value is the mean value of the six most sensitive paths to damage located at D1. Figure 8(b) represents the resultant DIs for all measurements.

- The pattern analysis (M3) applies covariance based-DI between all DI values in a damaged measurement with DI values in the baseline measurement, as shown in figure 7. The bigger the reference damage, the more significant the difference at some paths. Figure 8(c) represents the resultant values for all measurements.

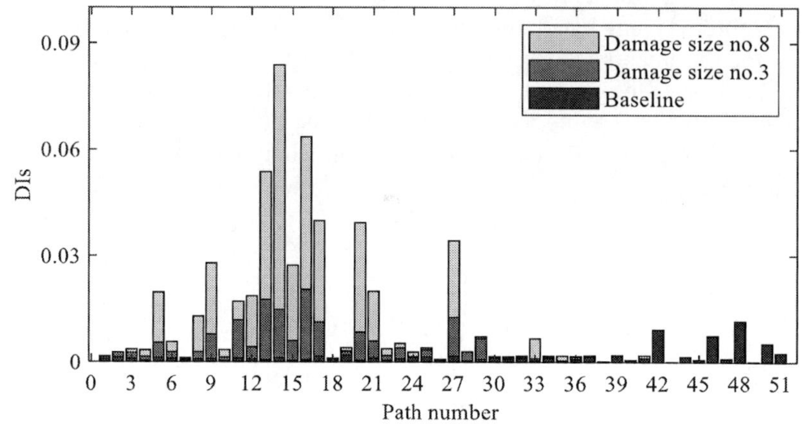

Figure (7). Pattern analysis. Bar plot of DIs of damage size no. 8 vs. damage size no. 3 vs. no damage

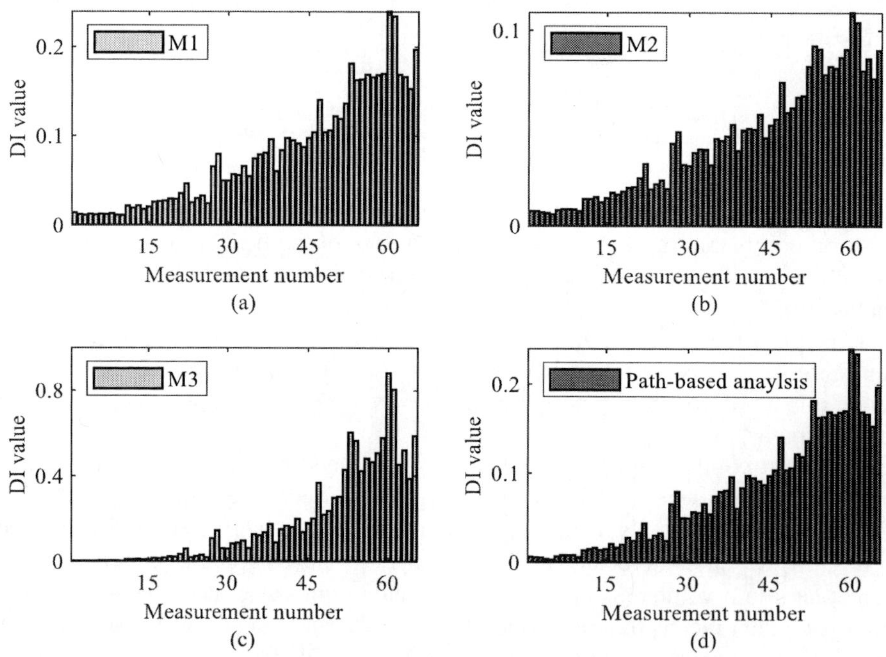

Figure (8). (a) Bar plot of M1 (b) Bar plot of M2 (c) Bar plot of M3 (d) Bar plot of a single path-based analysis

A null hypothesis test using Anderson Darling was carried out to test whether the DIs values of undamaged measurements for the 3 combined approaches and the single path-based analysis follow a normal distribution or not. M1, M2, M3, and the single path-based analysis are normally distributed. The PFA was set to 1%, which allows the threshold value to be estimated for each approach and the single path-based analysis using equation (1). The estimated threshold values for damage located at D1 are presented in the table II.

Table II. Threshold values

Approaches	M1	M2	M3	$T_3 - T_8$	$T_2 - T_8$	$T_4 - T_9$
$\hat{a}_{threshold}$	0.019	0.0098	0.0034	0.0013	0.00080	0.0015

4.1 POD curve generation

The generated POD curves are shown in figure 9. The solid line in the figure is related to the POD curve with a confidence bound of 50%. Meanwhile, the dashed line is related to increasing the confidence bound to 95%.

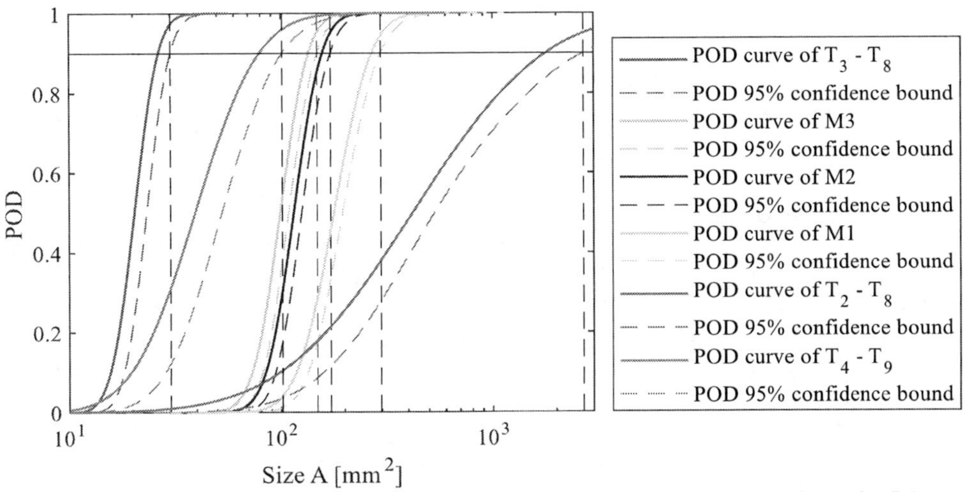

Figure (9). POD curves of 3 approaches and path-based analysis for damage located at D1

$a_{90|95}$ values can be obtained from the POD curves. It is the vertical dashed black lines as shown in figure 9. The values are presented in the following table III.

Table III. $a_{90|95}$ values

$a_{90\|95}$ [mm^2]			
Approaches	D1	D2	D3
M1	297	110	229
M2	172	81	135
M3	148	58	93
$T_3 - T_8$ path	30	56	112
$T_2 - T_8$ path	2682	23	435
$T_4 - T_9$ path	101	18212	46

The application of single path-based analysis highly depends on the choice of the best reference path. The results show that path [$T_3 - T_8$] can detect the smallest damage size for damage located at D1. That is not the case for D2 and D3 locations, [$T_2 - T_8$] and [$T_4 - T_9$] paths, respectively, can detect the smallest damage size. Also, the results show that the application of a path-based approach using path [$T_4 - T_9$] for damage located at D2 provides unrealistic results. The smallest damage size that can be detected is roughly $18212 mm^2$.

The application of the M1 approach provides the largest damage size that can be reliably detected, at the three damage locations. The $a_{90|95}$ values are the largest within this case study. The application of M2 and M3 approaches show slightly different results. The M2 approach is usually higher than M3 one, despite the different damage locations.

Using the M1 approach results in a significant decrease in the sensitivity of the SHM system used compared to a single path-based data analysis and other proposed approaches. The application of M2 and M3 approaches regarding damage location obtain similar results. The application of the single path-based anaylsis is the most sensitive approach in detecting the damage, however, the choice of the best reference path is highly dependent on the damage location.

6. Conclusion

Within this paper, three different approaches, pattern analysis, mean of maximum six paths, and maximum value analysis were proposed, and compared with a single path-based analysis. As the reference damage location changes, the generated POD curve and $a_{90|95}$ values vary.

The single path-based approach can detect the smallest damage size regarding the damage location. The sensitivity of single path-based data analysis is highly dependent on the choice of the best reference path. The application of M2 and M3 approaches show similar results for damage located at D1, D2, and D3. In contrast, the M1 approach can only detect more enormous damages reliably compared to the other approaches.

Statistically, combining approaches can quantify the detection probability of the whole guided wave SHM system when damage location is known or unknown. The application of M2 and M3 approaches can detect damage size without significant loss in sensitivity when assessing the whole SHM system for a known damage position.

7. References

[1] P. Cawley, "Structural health monitoring: Closing the gap between research and industrial deployment," *Struct. Heal. Monit.*, vol. 17, no. 5, pp. 1225–1244, 2018, doi: 10.1177/1475921717750047.

[2] D. W. Kelly, M. L. Scott, and R. S. Thomson, "Composite aircraft structures," *Model. Complex Eng. Struct.*, pp. 247–274, 2007, doi: 10.1061/9780784408506.ch09.

[3] E. Monaco, V. Memmolo, F. Ricci, N. D. Boffa, and L. Maio, "Guided waves based SHM systems for composites structural elements: statistical analyses finalized at probability of detection definition and assessment," *Heal. Monit. Struct. Biol. Syst. 2015*, vol. 9438, p. 94380M, 2015, doi: 10.1117/12.2084334.

[4] F. P. Grooteman, "Damage detection and probability of detection for a SHM system based on optical fibres applied to a stiffened composite panel," *Int. Conf. Noise Vib. Eng. 2012, ISMA 2012, Incl. USD 2012 Int. Conf. Uncertain. Struct. Dyn.*, vol. 4, pp. 3317–3330, 2012.

[5] C. Mandache, M. Genest, M. Khan, and N. Mrad, "Considerations on Structural Health

Monitoring Reliability," *CINDE Int. Work. Comference*, no. November, 2011, [Online]. Available: http://212.8.206.21/events/NDTCanada2011/proceedings/papers/17_Mandache_Rev1.pdf.

[6] V. K. Rentala, P. Mylavarapu, and J. P. Gautam, "Issues in estimating probability of detection of NDT techniques – A model assisted approach," *Ultrasonics*, vol. 87, pp. 59–70, 2018, doi: 10.1016/j.ultras.2018.02.012.

[7] J. Moll *et al.*, "Guided waves for damage detection in complex composite structures: The influence of omega stringer and different reference damage size," *Appl. Sci.*, vol. 10, no. 9, pp. 1–20, 2020, doi: 10.3390/app10093068.

[8] C. R. Farrar and K. Worden, "An Introduction to Structural Health Monitoring," in *New Trends in Vibration Based Structural Health Monitoring*, A. Deraemaeker and K. Worden, Eds. Vienna: Springer Vienna, 2010, pp. 1–17.

[9] C. Annis, E. Bray, H. Hardy, and P. M.-S. Hoppe, "Nondestructive evaluation system reliability assessment," *United States Dep. Defense, Wright-Patterson AFB, Handb. MIL-HDBK-1823A*, 2009.

[10] G. A. Georgiou, "PoD curves, their derivation, applications and limitations," *Insight Non-Destructive Test. Cond. Monit.*, vol. 49, no. 7, pp. 409–414, 2007, doi: 10.1784/insi.2007.49.7.409.

[11] R. Zhang and S. Mahadevan, "Fatigue Reliability Analysis Using Nondestructive Inspection," *J. Struct. Eng.*, vol. 127, no. 8, pp. 957–965, 2001, doi: 10.1061/(asce)0733-9445(2001)127:8(957).

[12] I. Mueller and C. P. Fritzen, "Inspection of piezoceramic transducers used for structural health monitoring," *Materials (Basel).*, vol. 10, no. 1, 2017, doi: 10.3390/ma10010071.

[13] A. P. Berens, "NDE reliability data analysis," *ASM Handbook.*, vol. 17, pp. 689–701, 1989.

[14] S. Torkamani, S. Roy, M. E. Barkey, E. Sazonov, S. Burkett, and S. Kotru, "A novel damage index for damage identification using guided waves with application in laminated composites," *Smart Mater. Struct.*, vol. 23, no. 9, 2014, doi: 10.1088/0964-1726/23/9/095015.

[15] J. B. Ihn, L. Pado, M. S. Leonard, M. P. Desimio, and S. E. Olson, "Development and performance quantification of an ultrasonic structural health monitoring system for monitoring fatigue cracks on a complex aircraft structure," *Struct. Heal. Monit. 2011 Cond. Maint. Intell. Struct. - Proc. 8th Int. Work. Struct. Heal. Monit.*, vol. 2, pp. 1899–1906, 2011.

[16] C. M. Schubert Kabban, B. M. Greenwell, M. P. DeSimio, and M. M. Derriso, "The probability of detection for structural health monitoring systems: Repeated measures data," *Struct. Heal. Monit.*, vol. 14, no. 3, pp. 252–264, 2015, doi: 10.1177/1475921714566530.

[17] M. Moix-Bonet, P. Wierach, R. Loendersloot, and M. Bach, "Damage assessment in composite structures based on acousto-ultrasonics—evaluation of performance," in *Smart Intelligent Aircraft Structures (SARISTU)*, Springer, 2016, pp. 617–629.

Phase Inversion for Accurate Extraction of the Harmonic Thermal Response in Active Infrared Thermographic NDT

Saeid HEDAYATRASA [1,2] Gaétan POELMAN [1], Mathias KERSEMANS [1]

[1] Mechanics of Materials and Structures (UGent-MMS), Department of Materials, Textiles and Chemical Engineering (MaTCh), Ghent University, Technologiepark-Zwijnaarde 46, 9052 Zwijnaarde, Belgium;
Phone: +32 9 3310421, email: Saeid.Hedayatrasa@UGent.be, Gaetan.Poelman@UGent.be
Mathias.Kersemans@UGent.be
[2] SIM Program M3 DETECT-IV, Technologiepark-Zwijnaarde 48, 9052 Zwijnaarde, Belgium;
email: Saeid.Hedayatrasa@UGent.be

Abstract
Infrared thermography is an efficient non-destructive testing technique in which heat waves are actively stimulated in the test piece by an external device, and an infrared camera measures the evolution of the surface temperature for revealing hidden defects as localized anomalies. The excitation amplitude is often sinusoidally modulated for a few cycles to obtain an improved detectability and to tune the detection depth. However, the commonly used excitation sources are heating-only (i.e. mono-polar) and thus constantly heat up the sample during the excitation, which leads to an undesired ascending trend in the measured thermal signal. This paper experimentally demonstrates the concept of phase inversion in infrared thermography as a physics-based technique for filtering this ascending trend (i.e. DC component) and extracting the desired harmonic (i.e. AC) component of thermal signal. The technique is validated for optical infrared thermography and low-power vibrothermography of a carbon fiber reinforced polymer aerospace component with impact damages.

Keywords: Non-destructive testing (NDT), Optical infrared thermography, Vibrothermography, Composite, Harmonic thermal response

1. Introduction

Infrared thermography is an efficient non-destructive testing (NDT) technique for fast full-field inspection of large components [1]. The heat wave is actively stimulated using an external device and an infrared camera measures the evolution of the surface temperature. The recorded surface temperature is then evaluated and post-processed for detection and quantification of internal defects. The excitation device may be an optical lamp (optical thermography), a vibrational actuator (vibrothermography) or an electromagnetic coil (eddy current thermography).

In order to tune the probing depth range of the stimulated heat wave and to increase the signal-to-noise ratio of the thermal signal, the excitation signal is often sinusoidally modulated for a few cycles at a constant frequency (i.e. lock-in thermography) [2-4], is swept over a given frequency range or, in a more complex form, is frequency/phase modulated [5-13]. However, all commonly used excitation sources of infrared thermography are heating-only, and as a result, constantly heat up the sample while their amplitude is modulated from zero to a maximum level. This leads to an ascending trend in the measured thermal response which contains very low frequency components that are irrelevant to the desired modulated response. Therefore, it is common practice to remove this ascending trend (i.e. DC component) by subtraction of low order polynomials [14-17]. The filtered results anyway highly depend on the quality of the original thermal dataset and the applied polynomial function.

This paper experimentally demonstrates the concept of phase inversion in lock-in infrared thermography for robust and stable extraction of the desired harmonic (i.e. AC) component of the thermal response. The technique is applied to optical infrared thermography and low-power vibrothermography of a carbon fiber reinforced polymer (CFRP) aerospace component with impact damages, and its good performance is demonstrated.

2. Materials and Methods

The inspected CFRP component is a stiffened panel from the vertical stabilizer of an Airbus A320 aircraft (see Figure 1). Three barely visible impact damages (BVIDs) were created by impacting the front side of the component with a 7.1 kg drop weight from three different drop heights (35 cm, 20 cm and 30 cm for $BVID_1$, $BVID_2$ and $BVID_3$, respectively). The component is inspected from the front side (Figure 1(a)) through (i) optical thermography in reflection mode and (ii) low-power vibrothermography.

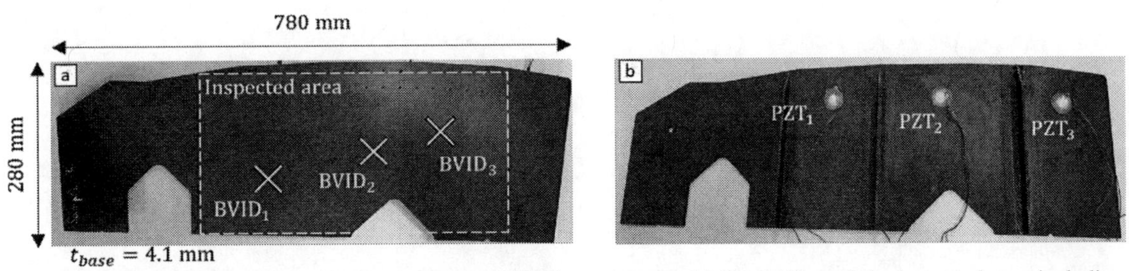

Figure 1. Inspected carbon fibre reinforce polymer (CFRP) component: (a) Front side and the inspected area including three barely visible impact damages (BVIDs), and (b) the stiffened backside on which three low-power piezoelectric transducers (PZTs) are glued for vibrational excitation.

A synchronized thermographic system (edevis® GmbH) was used and the surface temperature was measured by a FLIR A6750sc infrared camera at a sampling rate of 10 Hz. The camera has a cryo-cooled InSb detector which operates in the spectral range of 3-5 μm and has a pixel density of 640 × 512, a noise equivalent differential temperature (NEDT) of < 20 mK and a bit depth of 14 bit. The output of the infrared camera is given in digital level (DL) scale, which corresponds to the intensity of the recorded infrared radiation. For optical thermography, the front side was illuminated using two Hedler H25s halogen lamps at a stand-off distance of ~1000 mm. Each lamp has a nominal power of 2.0 kW and a color temperature of 3200 K, and is covered with two PMMA plates to filter out the interfering infrared spectrum of the excitation source. The lamp's signal generator has an output voltage range of 0 to 10 V, and the excitation signal is set to an amplitude of 9 V. For vibrothermography, three low-power piezoelectric transducers (PZTs) type EPZ-27MS44W from Ekulit, with a diameter of 27 mm were glued to the backside of the CFRP coupon (using Phenyl salicylate) as shown in Figure 1(b). A PXIe-5413 arbitrary waveform generator together with a Falco System WMA-300 voltage amplifier were then used to supply an excitation voltage of 300 Vpp to the three PZTs in parallel.
Both optical thermography and vibrothermography experiments were performed with 3 cycles of lock-in excitation. The optical excitation was performed at a modulation frequency of 0.1 Hz, and the vibrational excitation was applied at a frequency of 27 kHz and its amplitude was modulated with a

frequency of 0.03 Hz. The vibrational excitation frequency of 27 kHz, which provides a good detectability for all three BVIDs, is obtained from an initial vibrothermographic spectroscopy experiment using two time-reversed broadband vibrational excitations according to [6].

In order to investigate the concept of phase inversion in thermography, the experiments were performed first by applying the original excitation (i.e. voltage) signal S_0, and then by applying its phase-inversed one S_π (Figure 2(a)). The latter is performed after letting the sample return to the steady-state condition. In theory, and as confirmed by the following experimental results, subtraction of the two thermal responses, removes the DC component of the thermal signal and provides the desired AC component corresponding to the theoretical bi-polar excitation signal S_0^{AC} shown in Figure 2(b). Moreover, it can be shown that the proposed phase inversion technique also removes nonlinearity of thermal response with respect to the applied voltage signal. This is due to the invariance of even order nonlinearities (e.g. the quadratic nonlinearity induced by ohmic heating) to the phase inversion.

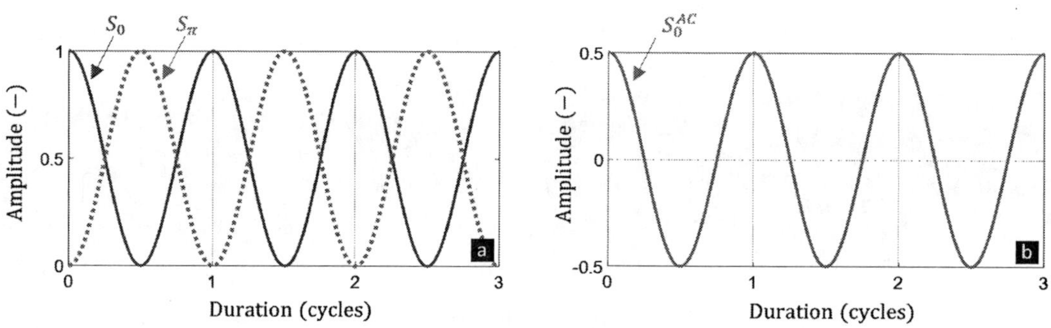

Figure 2. The 3-cycle lock-in excitation signals applied for both optical thermography and vibrothermography experiments: (a) the original excitation signal S_0 and its phase-inversed one S_π and (b) its theoretical bi-polar (AC) excitation signal S_0^{AC}

The measured thermal response also includes an initial cold stage prior to the excitation and a cooling down stage after the end of excitation. The mean value of the initial cold stage is used for pixel-wise cold image subtraction of the thermal dataset, and the cooling down stage is intended to capture the latency of the sample's thermal response to the applied excitation.

3. Phase Inversion in Active Infrared Thermography

3.1. Optical thermography

The performance of phase inversion in optical thermographic inspection of the CFRP panel is presented in Figure 3. The thermal responses (T_0 and T_π) to the two phase-inversed excitation signals (S_0 and S_π) at an arbitrary pixel is shown in Figure 3(a), after cold image subtraction.

According to the results: the DC component of the thermal response is calculated by summation (Figure 3(a)) and the AC component is calculated by subtraction of the two thermal responses (Figure 3(b)). Moreover, there is a minor higher harmonic (i.e. with a frequency twice of the modulation frequency) present on the calculated DC response which can be explained by e.g. the nonlinear relation of the lamp's power to the excitation voltage signal. These results provide evidence for the efficiency of the phase inversion technique for both extracting the AC component and filtering the prominent nonlinearities of the thermal response.

In order to evaluate the influence of the DC removal on defect detectability, the thermal dataset is post-processed with the fast Fourier transform (FFT) and the corresponding phase maps (at 0.1 Hz) are calculated. By comparing the phase maps of the raw thermal signal T_0 (Figure 3(c)) and the extracted AC component T_0^{AC} (Figure 3(d)) a significant difference in defect detectability is observed. The raw signal reveals the three BVIDs and the backside stiffeners, and also provides a minor indication of the PZTs glued to the backside of panel. However, the extracted AC response provides an exclusive indication of the BVIDs. This is because of the spectral leakage of the relatively low-frequency thermal responses (contained in the DC response) which have a deeper probing depth than the applied modulation frequency. These results clearly demonstrate the added value of phase inversion in infrared thermographic inspection of materials.

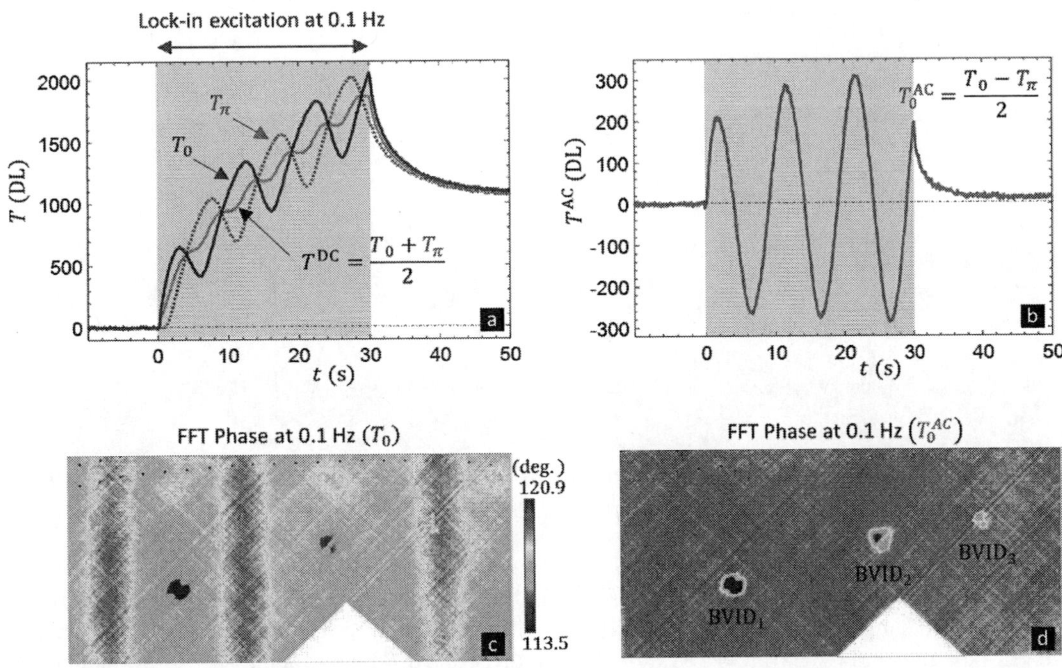

Figure 3. (a) Optical thermography thermal responses T_0 and T_π at an arbitrary pixel, and (b) the extracted AC component T_0^{AC}. Phase maps at $f = 0.1$ Hz of (c) the thermal response T_0 and (d) the extracted AC component T_0^{AC}.

3.1. Vibrothermography

The performance of phase inversion is further investigated for low-power vibrothermographic inspection of the CFRP panel. In vibrothermography, the inspection surface is generally cold and only the defected areas pop up as subsurface heat sources. The main challenge is the relatively low signal-to-noise ratio of vibrothermographic inspection results. Moreover, there is a latency in measuring the

thermal response of a defect depending on its depth. Therefore, it is quite challenging (if not impossible) to properly extract the AC component using classical polynomial subtraction method.
Instead, applying the phase inversion method to extract the AC component of the vibrothermal response leads to good results, as demonstrated in Figure 4. It can be seen that besides the BIVDs, the three PZTs are also detected as heat sources due to their self-heating during the excitation.

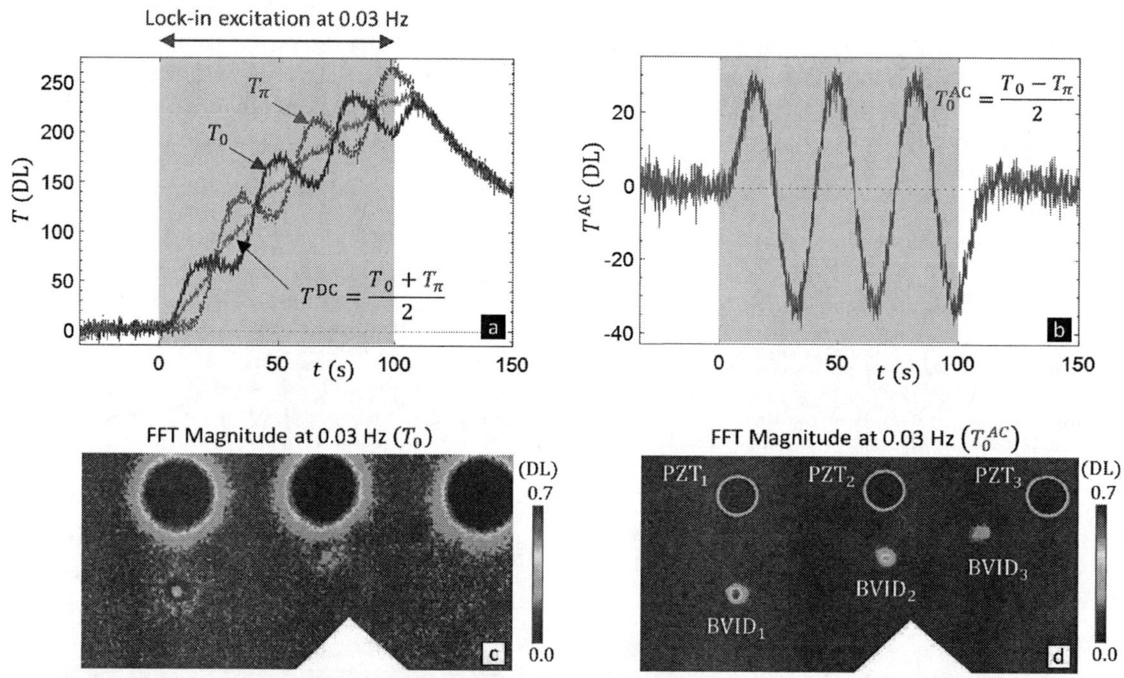

Figure 4. (a) Vibrothermal responses T_0 and T_π at a pixel on PZT_2 , and (b) the extracted AC component T_0^{AC}. Magnitude maps at f = 0.03 Hz of (c) the vibrothermal response T_0 and (d) the extracted AC component T_0^{AC}.

Due to the low thermal signal measured at the BVIDs, an arbitrary pixel on PZT_2 is selected to illustrate the effectiveness of phase inversion in the separation of the AC component (Fig (a,b)). The significant delay of the thermal signal of the PZT after diffusion through the entire thickness of the component is clear in both the raw and the extracted AC signals (Figure 4(a,b)). The extracted DC component of the thermal signal shows a minor indication of nonlinearities filtered from the thermal signal, which may be explained by the self-heating nonlinearity of the PZT with respect to the excitation voltage. Due to the damping induced by heat diffusion through the thickness, the nonlinearity measured at the inspection surface has a very low amplitude.

In the case of vibrothermography, the FFT magnitude maps are preferred which provide a good indication of subsurface heat sources on the cold inspection surface. The magnitude map of the raw thermal signal (Figure 4(c)) shows a very poor detectability of the BVIDs and it is predominantly affected by the in-plane heat diffusion induced by self-heating of PZTs. In contrary, the magnitude map of the extracted AC component (Figure 4(d)) provides a distinctive indication of all three BVIDs and also an improved indication of PZTs in which the effect of lateral heat diffusion is suppressed.

3. Conclusions

Phase inversion in active infrared thermography is proposed to accurately decouple the AC component from the measured thermal signal, and as such to increase the defect detectability. The method has been demonstrated on an aircraft CFRP component with backside stiffeners and impact damages, for both optical thermography and low-power vibrothermography inspection. It was shown that the phase inversion method leads to improved SNR and increased defect detectability.

Acknowledgements

The authors acknowledge the SBO project DETECT-IV (Grant no. 160455), which fits in the SIM research program MacroModelMat (M3) coordinated by Siemens (Siemens Digital Industries Software, Belgium) and funded by SIM (Strategic Initiative Materials in Flanders) and VLAIO (Flemish government agency Flanders Innovation & Entrepreneurship). The authors also acknowledge Fonds voor Wetenschappelijk Onderzoek-Vlaanderen (FWO-Vlaanderen) through grant 1S11520N. The authors further thank SABCA Limburg for providing the aircraft panel.

References

1. F. Khodayar, S. Sojasi, and X. Maldague, 'Infrared thermography and NDT: 2050 horizon', Quantitative InfraRed Thermography Journal, Vol 13, No 2: p. 210-231, 2016.
2. J. Liu, W. Yang, and J. Dai, 'Research on thermal wave processing of lock-in thermography based on analyzing image sequences for NDT', Infrared Physics & Technology, Vol 53, No 5: p. 348-357, 2010.
3. J. Segers, et al., 'In-plane local defect resonances for efficient vibrothermography of impacted carbon fiber-reinforced polymers (CFRP)', NDT & E International, Vol 102: p. 218-225, 2019.
4. M. Razani, A. Parkhimchyk, and N. Tabatabaei, 'Lock-in thermography using a cellphone attachment infrared camera', Aip Advances, Vol 8, No 3: p. 035305, 2018.
5. K. Chatterjee, et al., 'A comparison of the pulsed, lock-in and frequency modulated thermography nondestructive evaluation techniques', NDT & E International, Vol 44, No 7: p. 655-667, 2011.
6. S. Hedayatrasa, et al., 'Vibrothermographic spectroscopy with thermal latency compensation for effective identification of local defect resonance frequencies of a CFRP with BVID', NDT & E International, Vol 109: p. 102179, 2020.
7. N. Tabatabaei, 'Matched-Filter Thermography', Applied Sciences, Vol 8, No 4: p. 581, 2018.
8. S. Hedayatrasa, et al., 'Vibro-Thermal Wave Radar: Application of Barker coded amplitude modulation for enhanced low-power vibrothermographic inspection of composites', Materials, Vol 14, No 9: p. 2436, 2021.
9. S. Hedayatrasa, et al., 'Performance of frequency and/or phase modulated excitation waveforms for optical infrared thermography of CFRPs through thermal wave radar: A simulation study', Composite Structures, Vol 225: p. 111177, 2019.
10. X. Lu, Q. Yi, and G. Tian, 'A Comparison of Feature Extraction Techniques for Delamination of CFRP Using Eddy Current Pulse-Compression Thermography', IEEE Sensors Journal, Vol 20, No 20: p. 12415-12422, 2020.

11. J. Liu, et al., 'Study of inspection on metal sheet with subsurface defects using linear frequency modulated ultrasound excitation thermal-wave imaging (LFM-UTWI)', Infrared Physics & Technology, Vol 62: p. 136-142, 2014.
12. A. Rani, R. Mulaveesala, and V. Kher, 'An analytical approach for frequency modulated thermal wave imaging for testing and evaluation of glass fiber reinforced polymers', IOP SciNotes, 2021.
13. Q. Shi, et al., 'Barker-coded Modulation Laser Thermography for CFRP Laminates Delamination Detection', Infrared Physics & Technology, Vol 98: p. 55-61, 2019.
14. J. Gong, et al., 'Investigation of carbon fiber reinforced polymer (CFRP) sheet with subsurface defects inspection using thermal-wave radar imaging (TWRI) based on the multi-transform technique', NDT & E International, Vol 62: p. 130-136, 2014.
15. G. Silipigni, et al., 'Optimization of the pulse-compression technique applied to the infrared thermography nondestructive evaluation', NDT & E International, Vol 87: p. 100-110, 2017.
16. K. Chatterjee and S. Tuli, 'Image enhancement in transient lock-in thermography through time series reconstruction and spatial slope correction', IEEE Transactions on Instrumentation and Measurement, Vol 61, No 4: p. 1079-1089, 2011.
17. S. Hedayatrasa, et al., 'On the application of an optimized Frequency-Phase Modulated waveform for enhanced infrared thermal wave radar imaging of composites', Optics and Lasers in Engineering, Vol 138, 2021.

A Numerical Modeling Tool for Scanning Air-Coupled Ultrasonic Testing Based on the Elastodynamic Finite Integration Technique (EFIT)

Frank SCHUBERT [1], Martin BARTH [1]

[1] Systems for Testing & Analysis, Fraunhofer Institute for Ceramic Technologies and Systems IKTS, Dresden, Germany; Phone: +49 351 88815-523, Fax +49 351 88815-509; e-mail: frank.schubert@ikts.fraunhofer.de, martin.barth@ikts.fraunhofer.de

Abstract
In this article, a numerical simulation tool for the calculation and optimization of scanning air-coupled ultrasonic testing (ACUT) of nearly arbitrary multilayer structures with and without flaws based on the elastodynamic finite integration technique (EFIT) is presented. It includes 2-D and 3-D solvers for flat and curved isotropic and anisotropic layered structures in transmission, pulse-echo, and pitch-catch configuration. During the development, special attention was paid to the minimization of computing time in the scanning mode, which was achieved by various measures. The paper presents the freely configurable parameters of the EFIT solver, typical result output, special features, and an optional LabVIEW-based user interface. Along the way, several simulation examples from air-coupled ultrasonic testing are demonstrated.

Keywords: Scanning air-coupled ultrasonic testing, ACUT, numerical simulation, elastodynamic finite integration technique, EFIT

1. Introduction

For some time now, air-coupled ultrasonic testing is again in the focus of users, developers, and device manufacturers due to its non-contact operation mode. Although it follows the same physical principles than conventional ultrasonic testing, it has a number of special features that must be taken into account when designing measurement setups and evaluating measurement data. Crucial aspects are the comparatively small test frequencies and the large differences in acoustic impedance and speed of sound between air and the solid medium under test. In the numerical simulation of conventional ultrasonic testing, the propagation of sound in the surrounding air is usually neglected in order to minimize the computational effort. To calculate air-coupled ultrasound on the other hand, the air must be explicitly taken into account, which brings new challenges for the simulation process, especially when realizing scanning B- and C-Scan operations.

In the present work we used the well-established elastodynamic finite integration technique (EFIT) as basis for our solver due to its flexibility and robustness [1-2]. Because of its explicit time-domain formalism on a staggered spatio-temporal grid, it produces stable and reliable results even for the typical large impedance mismatches in ACUT applications. During the development and the subsequent testing phase we never encountered any numerical problems or instabilities as sometimes reported from other numerical techniques. In the following chapter we will describe the basic free parameters of our software to demonstrate its scope of functions.

2. Program parameters

In this chapter the main free input parameters for the EFIT solver are described. They allow to set-up the ACUT model and to control the subsequent solver execution.

2.1 Model dimensions and matrix medium

In a first step, the overall dimensions of the computational model in an underlying 3-D Cartesian coordinate system (a cuboid) are defined. Additionally, the physical time duration of a single simulation is specified (later on we will see that multiple simulations can be combined to full B- and C-Scans). After that the fluid matrix medium surrounding the remaining solid parts of the measurement set-up is given. In principle, various fluids can be defined but in the given context the medium air is used. It is specified by a constant pressure wave speed (e.g. $c_P = 343$ m/s) and a mass density (e.g. $\rho = 1$ kg/m³) assuming an idealized fluid with zero shear stresses, i.e. the shear wave speed vanishes, $c_S = 0$ m/s). In the current version of the program, the absorption in air is neglected which is a sufficient approximation for low frequencies and short distances between transducers and specimen but insufficient at larger frequencies and distances. It is planned to extend the model with a realistic absorption mechanism in the future.

2.2 Layer set-up and additional scatterers

In a further step, an arbitrary sequence of solid and fluid layers embedded in the surrounding air can be defined. The single layers are described by their thickness, acoustic material properties (pressure and shear wave speed plus mass density) and by their lateral dimension within the model. For solid layers, an additional option for anisotropic material parameters is available. Currently, the anisotropy is restricted to transversely isotropic media like CFRP or GFRP. The standard geometry of the multilayer set-up is a flat plate, but if necessary, it can be bended to an arbitrary extent. In addition to the pure layers, additional scatterers and local indentations can be added to describe various types of defects and local irregularities in the plate topography. All scatterers are defined by their individual size, location, orientation, and material properties. They can be separately activated and deactivated in order to investigate their single scattering behaviour but also the interaction among themselves. Currently, ellipsoidal, cuboidal, and cylindrical shapes are supported but more complex geometries are also possible.

2.3 Transmitter and receiver configurations

In the model, transmitter and receiver can be defined separately so that various measurement set-ups are possible (pulse-echo, through-transmission, pitch-catch). The transducers are specified by their position, size, orientation, and focal length. Also unfocused transducers are possible. Moreover, additional shielding covers can be added in order to suppress direct acoustic crosstalk between transmitter and receiver (especially in pitch-catch configurations).

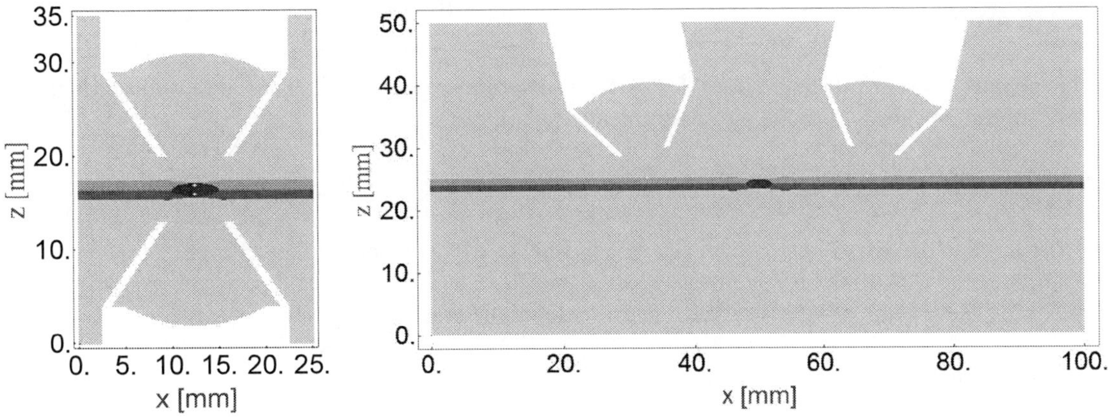

Figure 1. Cross-sectional view of a through-transmission (on the left) and a pitch-catch configuration (on the right) of spot weld testing incorporating a 2-sheet aluminum plate with weld nugget (black ellipse) and corresponding plate indentations caused by the welding tongs. Both transducers are acoustically shielded by an adequate cover.

In Fig. 1, two typical ACUT configurations (through-transmission and pitch-catch) for the inspection of spot weldings are displayed. For a pulse-echo configuration, transmitter and receiver are defined with identical properties. However, this pulse-echo set-up is barely used in ACUT practice.

Beside its geometrical properties, also the excitation signal of the transmitter can be specified. For this purpose, the time history of the input pulse, its amplitude, center frequency, and bandwidth can be defined. For most transducers used in ACUT, a rather limited bandwidth can be observed resulting in long pulses with a large number of cycles.

2.4 Program execution

Beside the model details, also the program execution can be influenced by the user. Normally, the spatio-temporal numerical grid is automatically determined by the software but the number of grid points per minimum wavelength can be specified for each of the three Cartesian axes separately. In this context, the spatial grid parameters are not only determined by the minimum wavelength but also by the size or thickness of the smallest geometrical feature specified in the model. For experienced users, it is possible to circumvent the automatic operation of the software by forcing a cubic grid for or by defining the three spatial grid parameters explicitly. The definition of the numerical grid parameters is completed by the dimension of the underlying EFIT solver (2-D or 3-D). The 3-D solver will model the real measurement set-up correctly but is usually characterized by a significantly larger computation time. The 2-D solver is much faster and already able to reproduce the most important acoustical phenomena in a certain cross-section of the 3-D model. However, due to its plane strain approximation, the results will differ quantitatively from the real 3-D case. Therefore, the 2-D solver is mostly used for testing and optimization and in cases where a qualitative result is sufficient. It can simply be replaced by the 3-D solver by setting a checkbox state.

3. Data output

In this chapter, the various data output files of the software are described. They will be generated during solver execution after the input parameters have been imported.

3.1 Model data

In order to visualize the ACUT model set-up and to check its correctness, the complete material allocation of the 3-D model is saved as a voxel data set. In addition, the central x/z cross-sectional plane of the model can be visualized as a grey-scale image (Fig. 1).

3.2 Surface topography

In addition to the distribution of material parameters, the surface topography of the complete layer set-up is also provided, again in terms of a 3-D voxel data set (Fig. 2). This representation is particularly useful if plate indentations and/or curved surfaces exist.

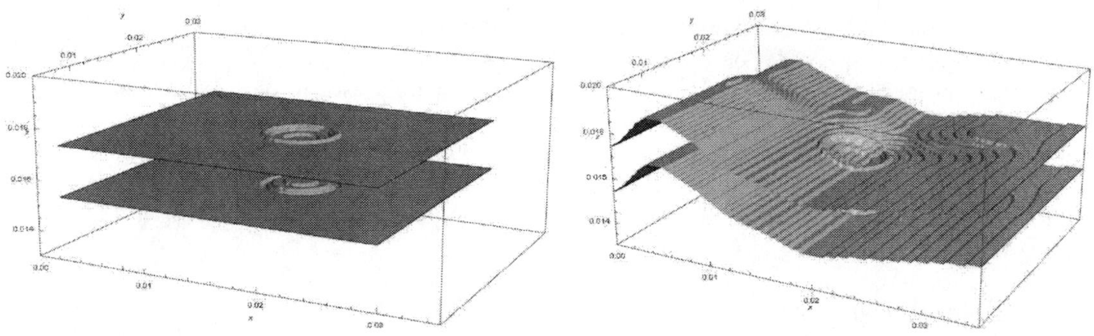

Figure 2. 3-D representations of the surfaces of the layer set-up. On the left: flat 2-sheet plate with local indentations caused by a spot weld tong. On the right: the same model, but with an overlaid curved surface.

3.3 Wavefront snapshots

The complete wave propagation process from the excitation of the input pulse at $t = 0$ until the pre-defined physical end time of the simulation can be visualized by wavefront snapshots as shown in Fig. 3. The start and end time of the sequence as well as the temporal distance between the single snapshots can be specified by the user with additional input parameters not mentioned in chapter 2. Moreover, the image section, the picture resolution, and the kind of field component to be visualized can be defined. Currently, the three vector components of the particle velocity as well as the absolute value of this vector are supported. In principle, also the pressure or the stress components in solid materials can be displayed. From the sequence of snapshots, animated movies of wave propagation can be generated.

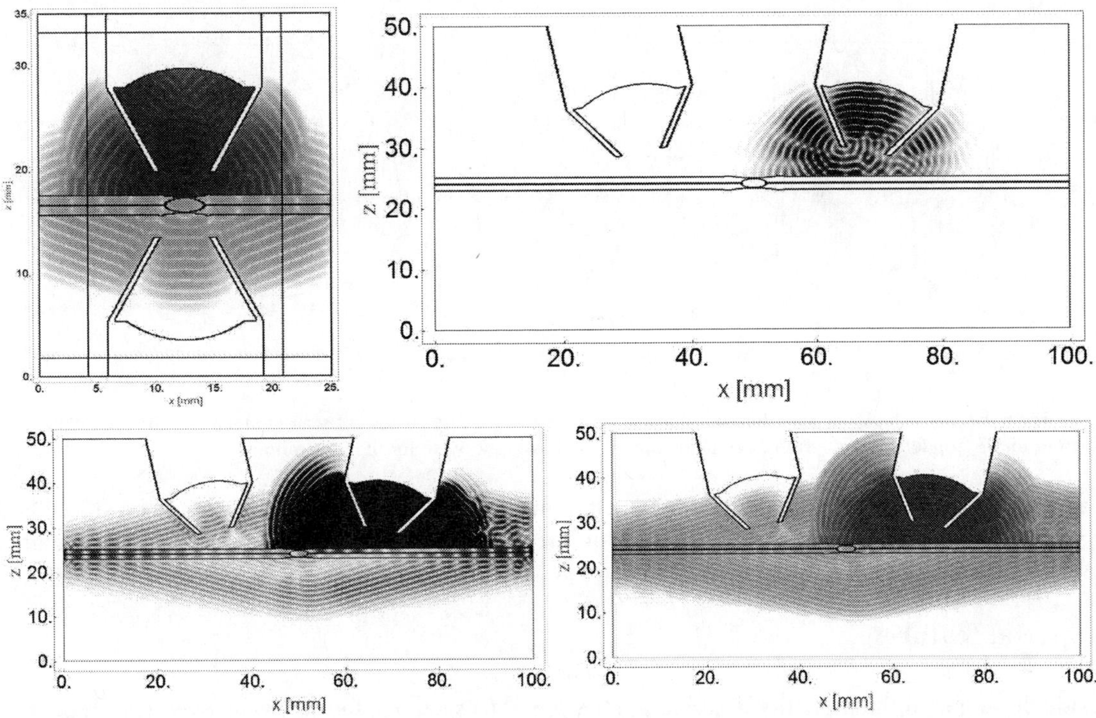

Figure 3. Typical grey scale images of air-coupled ultrasonic wave propagation. Top left: through-transmission inspection of a spot weld (logarithmic scale); top right: pitch-catch configuration of the same plate (linear scale); bottom left: later snapshot with transmitted field (linear scale); bottom right: the same snapshot in logarithmic scale. In the first picture, also the surrounding PML boundaries of the numerical model are displayed (see section 4.2).

Due to the usually large acoustic impedance mismatches between air and the specimen under test, large amplitude differences between transmitted and reflected wave fields can be observed. In order to avoid overdriven representations based on linear scales as shown in Fig. 3 (bottom left), logarithmic grey or colour scales can be used instead (Fig. 3, bottom right). In these pictures, all relevant details of the wave field can be properly observed, also the transmitted parts. As seen from Fig. 3, the EFIT simulations include transmitted and reflected waves in the air, guided waves in the solid layers as well as leaky waves in the air. The latter are detected by the second transducer on the left in order to determine guided wave distortions caused by the spot weld.

3.4 Time-domain signals

The most important output of the solver is the time-domain signal detected across the aperture of the receiving transducer. In this case, the time history of the pressure signal is calculated. In case of full B- or C-Scans (see section 4.1) the system response at each measuring point is stored as a sequence of time-domain signals (Fig. 4, on the left). From this raw data, the corresponding B- and C-Scans can be determined by appropriate time-domain windowing.

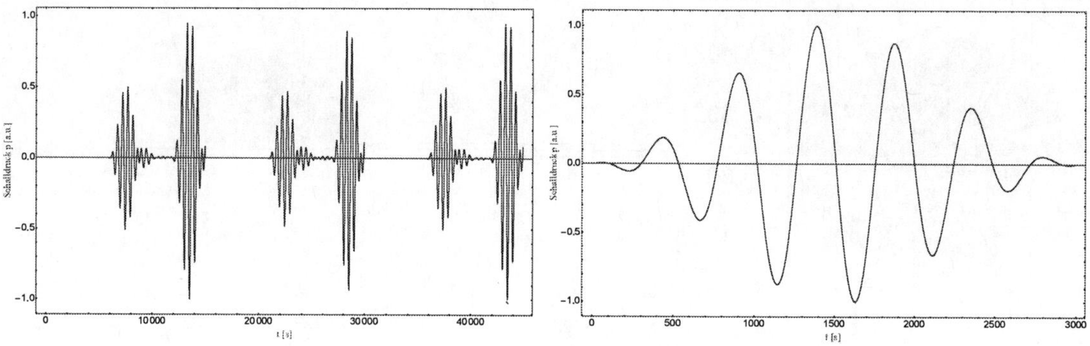

Figure 4. On the left: calculated time-domain signal at the receiver for three different scan points along the spot weld shown in Fig. 3 (top left). On the right: excitation signal used as input pulse for the transmitter.

In addition to the received signals, also the excitation signal used by the transmitter can be stored and displayed for documentation purposes (Fig. 4, on the right). It can be changed with regard to center frequency and bandwidth as already explained in section 2.3.

4. Special features

Beside the standard functionality described in the previous sections, the software provides a number of further powerful tools for the simulation of ACUT inspections. They are briefly explained in the following.

4.1 B- and C-Scans

In addition to single simulation runs with a fixed positions of the transducers relative to the inspected plate, a fully automated and smart scan mode is available in the program. For this purpose, a scan grid in the x/y plane across the top plate surface can be freely configured. After start of the solver, the software will automatically perform multiple simulation runs for all measuring points defined. During this process, several measures are taken to keep the computation time small and to avoid unnecessary multiple computations. In this context, the time period until the excited waves in the air start interacting with the layer set-up is only calculated once since this part of the wave field is constant for all measuring points. Moreover, scan points with a certain symmetry, i.e. an identical position relative to the plate and the scatterers inside, also need to be calculated only once. Finally, for pulse-echo and pitch-catch configurations, the surrounding air beneath the lower sheet can be neglected and replaced by a stress-free boundary between this layer and air (Fig. 5).

In the EFIT solver, the relative movement between plate and transducers during the scan is realized by a lateral shift of the plate while keeping the transducer configuration fixed (Fig. 6).

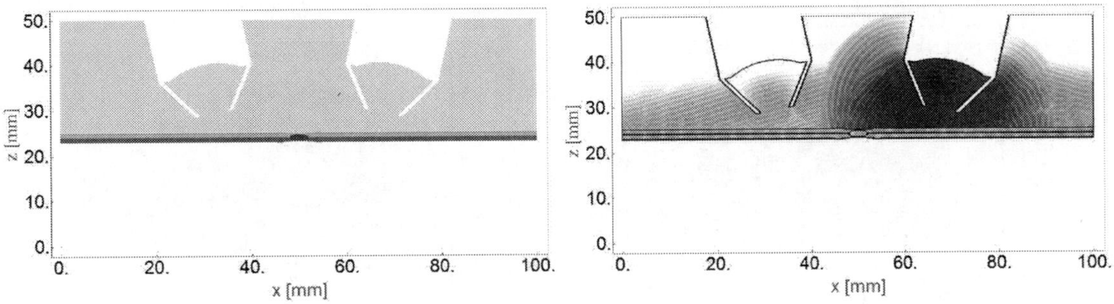

Figure 5. Neglect of the air beneath the lower sheet for a pitch-catch configuration (compare with Fig. 3, bottom right). In this case the sheet-air transition can be replaced by a stress-free boundary so that the transmitted wave field is omitted.

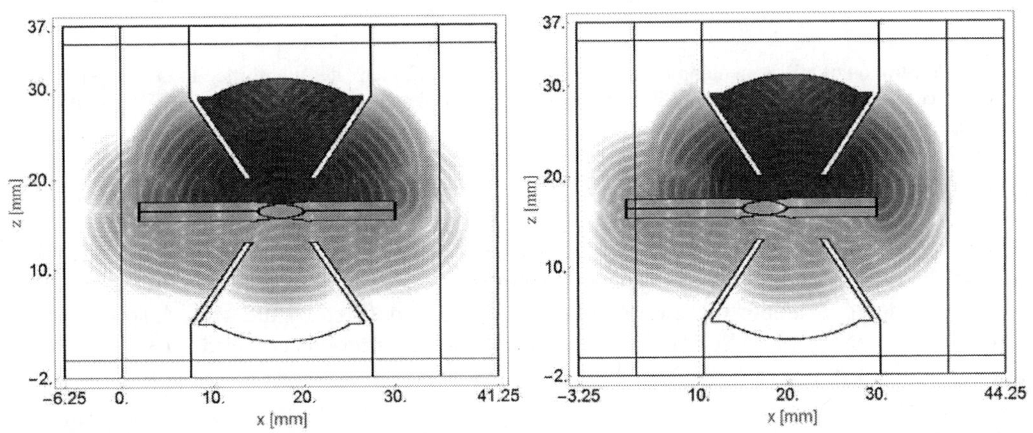

Figure 6. Exemplary demonstration of a B-Scan along the x-direction of a finite plate. It is shifted to the left while keeping the transducer configuration fixed. In the pictures, the surrounding PML zone is also shown (see section 4.2).

4.2 Perfectly matched layer boundaries and finite/infinite plates

In order to suppress disturbing reflections from the outer boundaries of the computational domain, effective perfectly matched layer (PML) boundaries [3] are incorporated in the EFIT model. With these absorptive boundaries, the computational domain (and thus, the computational effort) can be kept as small as possible, which is especially important for high-frequency 3-D simulations. The thickness of the PML zone can be defined for each of the six boundaries of the cuboidal computational domain separately. The absolute damping of the PML zones is controlled by only a single overall damping coefficient. The proper definition of this value ensures absolute numerical stability under all circumstances and was reached by implementing a new PML stability criterium that avoids unphysical group velocities being larger than the physical speed of sound. The details of this approach will be published elsewhere. As can be seen in Fig. 6 (on the right), the PML boundaries are compatible with laterally finite plates, even when the end points of the plate overlap with the PML region. In this context, also infinite and semi-infinite plates can be realized (Fig. 7).

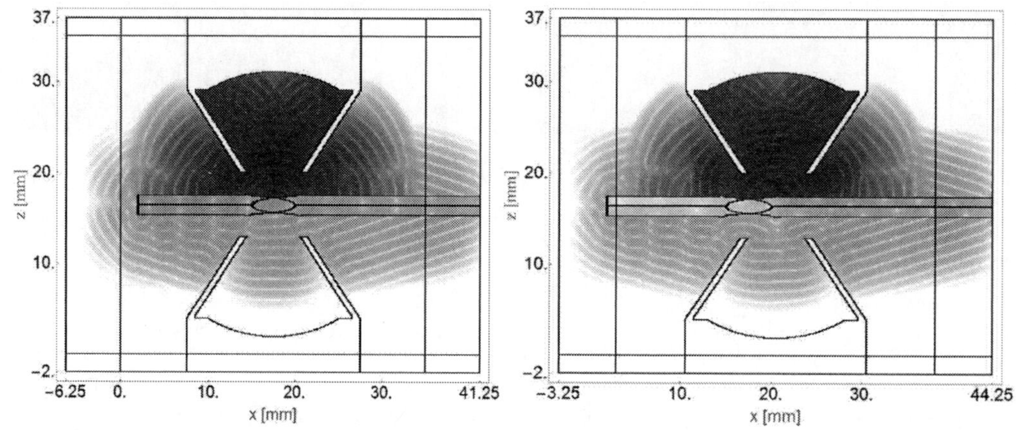

Figure 7. Demonstration of a B-Scan along the x-direction of a semi-infinite plate. It is shifted to the left while keeping the transducer configuration fixed. In the PML zone on the right, the plate is extrapolated to ensure its infinite extent in this direction.

4.3 Curved surfaces

As already demonstrated in section 3.2, the EFIT solver is also able to deal with (moderately) curved surfaces as shown in Fig. 2. For this purpose, the surface topography of the initially flat plate is modulated by adequate supporting points in the x/y plane as defined by the user. Also in this case, the general functionality of PML boundaries and the scanning mode is ensured for finite, semi-infinite, and infinite plates (Fig. 8 and 9).

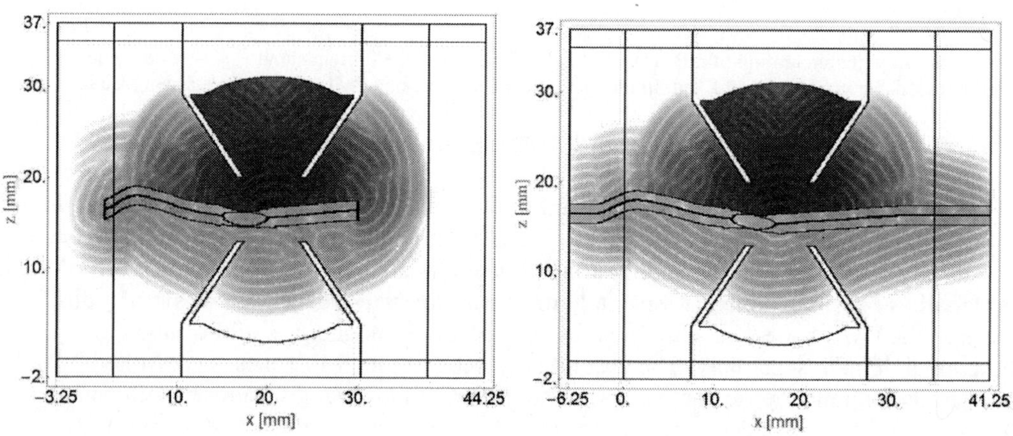

Figure 8. Demonstration of a scan along the x-direction of a curved finite plate (on the left) and a (partly) curved infinite plate (on the right).

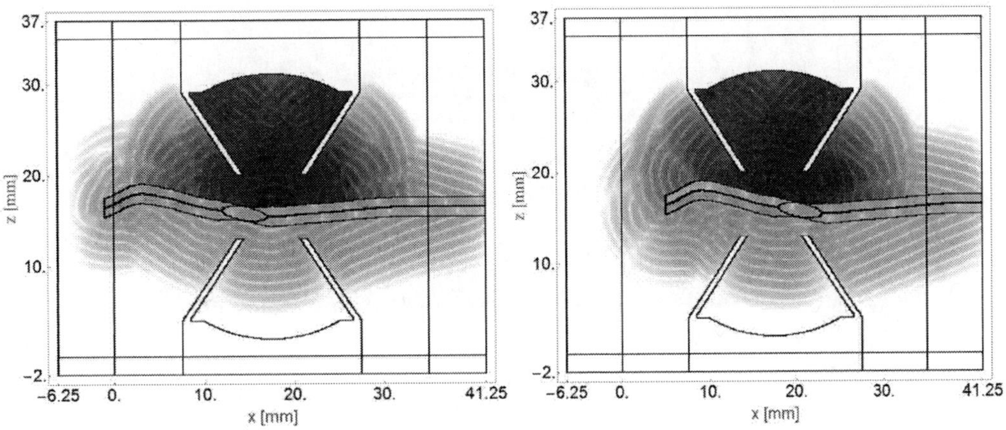

Figure 9. Demonstration of a scan along the x-direction of a (partly) curved semi-infinite plate. On the left: plate shifted to the left; on the right: plate shifted to the right.

5. Graphical user interface

The EFIT solver described in the previous chapters can be used as a stand-alone computation tool. It is fully controlled by a set of input parameters as described in chapters 2-4. These parameters are stored in a simple text file that is imported by the solver after start of execution. During the simulation, the solver generates several data output files that can be further processed and evaluated by external software. As an option, an appropriate graphical user interface (GUI) can also be delivered. It allows the definition of the model set-up in a graphical environment and can also be used to visualize and evaluate the numerical results. An example of a user-specific LabVIEW GUI is shown in Figs. 10 and 11.

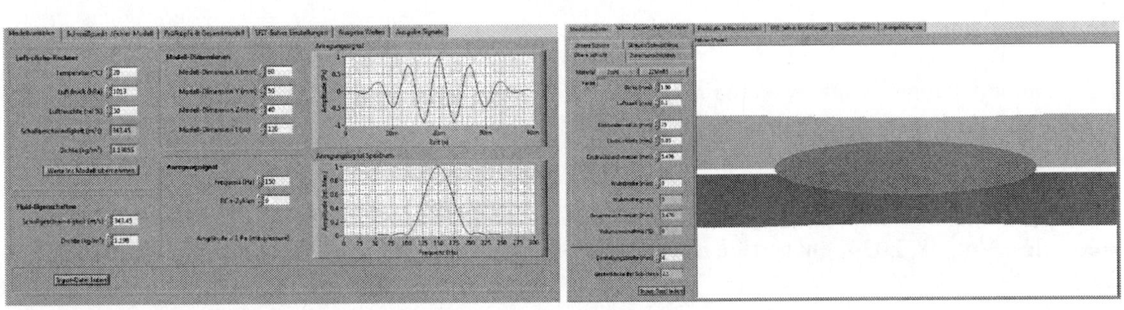

Figure 10. Examples from a LabVIEW GUI demonstrating the specification of the excitation signal (on the left) and the details of the spot weld geometry including weld nugget and weld tong indentation (on the right).

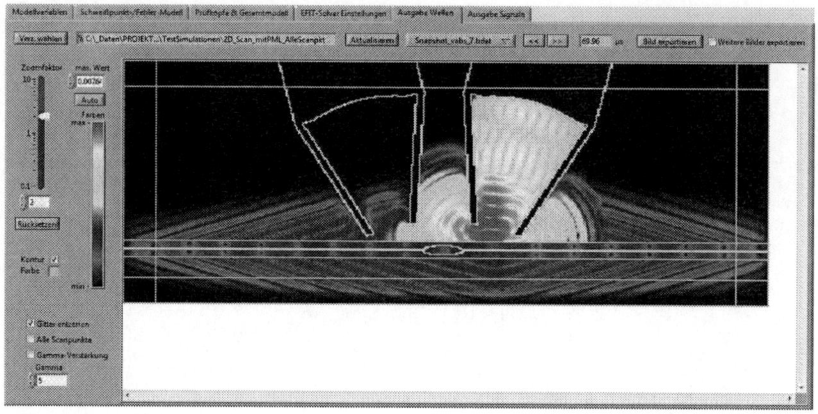

Figure 11. Example from a LabVIEW GUI demonstrating the coloured representation of a wavefront snapshot in a pitch-catch arrangement.

6. Conclusions

In this paper, a powerful numerical tool for 3-D air-coupled ultrasonic wave propagation problems based on the elastodynamic finite integration technique EFIT has been presented. It offers the definition of nearly arbitrary isotropic and anisotropic multilayer set-ups in pulse-echo, pitch-catch, and through-transmission configurations. Moreover, an unlimited number of various scatterers can be defined and placed inside or outside the layer stack. In addition, full B- and C-Scan functionality is provided and curved surfaces are supported as well. In the future, several extensions of the solver are planned, e.g. the incorporation of damping effects in air and the support of laser ultrasound applications where the standard ACUT transducers are replaced by thermoelastic sources and optical vibrometers or microphones, respectively.

References

1. F Schubert, 'Numerical time-domain modeling of linear and nonlinear ultrasonic wave propagation using finite integration techniques – Theory and applications', Ultrasonics 42, 221-229, 2004.

2. F. Schubert, T. Herzog, S. Walter, H. Heuer, 'Ultrasonic dual element probes revisited – Complex wave phenomena as a basis for transducer design and optimization', Proceedings of Meetings on Acoustics, Vol. 38, 2020, International Congress on Ultrasonics, Bruges, Belgium, 3-6 September 2019.

3. W. C. Chew, Q. H. Liu, 'Perfectly Matched Layers for Elastodynamics: A New Absorbing Boundary Condition', Journal of Computational Acoustics, 04(04), August 1997.

Ultrasonic Reconstruction of the Fiber Architecture of Composite Laminates

Xiaoyu YANG[1], Bing-feng JU[2], Mathias KERSEMANS[1]

[1] Mechanics of Materials and Structures (UGent-MMS), Ghent University, Gent, Belgium

Phone: +32 (0)9 331 04 09; e-mail: xiaoyu.yang@ugent.be

[2] The State Key Laboratory of Fluid Power Transmission and Control, Zhejiang University, Hangzhou, People's Republic of China

Abstract

By proper design of the fiber architecture, composites can be tailored to a specific application. However, due to the typical manufacturing process and possible impact events, there is a clear risk of introducing deviations of the intended fiber structure. Hence, it is of utmost importance to verify the intended fiber structure of the composite by means of non-destructive testing. This paper presents a planar ultrasound computed tomography (pU-CT) methodology for the 3D reconstruction of fiber structure in a composite laminate. The method relies on a classical ultrasonic pulse-echo scan and couples the analytic-signal response to a structure tensor process and a Gabor filter-based Information Diagram. The pU-CT is demonstrated on a 24-layer carbon fiber reinforced polymer laminate in order to reconstruct the ply-by-ply (i) thicknesses, (ii) out-of-plane fiber orientations, and (iii) in-plane fiber orientations. Robust 3D tomographic reconstruction results have been obtained, indicating the high potential of pU-CT as an alternative/complement for X-ray CT.

Keywords: ultrasonic pulse-echo technique, tomographic reconstruction, analytic-signal, fiber architecture, carbon fiber-reinforced polymers (CFRP)

1. Introduction

Carbon fiber reinforced polymers (CFRP) have been widely used in the aerospace and automobile industries. CFRP gets excellent performance by the right combination of fiber to resin and the designed orientation of fibers. However, the manufacturing of CFRP laminates often involves manual work, making it crucial to evaluate their structural quality. For example, in compression moulding, there is a clear risk to induce out-of-plane ply wrinkles having a considerable impact on the structural performance of the final composite part [1]. Hence, it is of utmost importance to verify the intended fiber structure of the manufactured composite by means of non-destructive testing.

Many researchers employ X-ray computed tomography (X-CT) to obtain detailed 3D information on the composite structures [2]. Unfortunately, X-CT also has some fundamental challenges: (i) poor resolution for large parts, (ii) health issues, (iii) low contrast for carbon composites, (iv) high cost, and (v) long scan time. Recently, researchers have proposed the use of ultrasound in the low MHz range, combined with the analytic-signal technique, for evaluating the fiber structure of a composite [3, 4].

The analytic-signal procedure coupled to the log-Gabor filter has shown high performance for robust interply tracking in CFRP laminates [5, 6]. Following the analytic-signal analysis, the out-of-plane ply orientation could be extracted by applying the structure tensor method to the instantaneous phase [4]. Besides, the in-plane fiber orientation is extracted by applying the Gabor filter-based Information Diagram (GF-ID) method to the instantaneous amplitude.

This paper presents a planar ultrasound computed tomography (pU-CT) method in view of reconstructing the local fiber architecture of multi-layer composites by bringing various processing techniques together. The pU-CT methodology is demonstrated on a CFRP laminate for the 3D reconstruction of fiber structure.

2. Planar Ultrasound Computed Tomography pU-CT methodology

The pU-CT methodology consists of three steps: (i) analytic-signal analysis to map the structural interfaces, (ii) structure tensor process to extract the out-of-plane ply orientation, and (iii) Gabor filter-based Information Diagram to extract in-plane fiber direction. The three steps are detailed in the following subsections.

2.1 Analytic-signal analysis to map the structural interfaces

An analytic-signal analysis is employed to distinguish between the ultrasonic reflections from different interfaces, i.e., front- and back surfaces, delaminations, and interplies. The analytic-signal $s_a(t)$ is obtained by applying the Hilbert transform to the acquired signal $s(t)$ as follows

$$s_a(t) = s(t) + ig(t), \quad (1)$$

$$g(t) = \int_{-\infty}^{+\infty} \frac{s(\tau)}{\pi(t-\tau)} d\tau. \quad (2)$$

The instantaneous amplitude $A_{inst}(t)$ and instantaneous phase $\phi_{inst}(t)$ are then derived

$$A_{inst}(t) = \sqrt{[Re(s_a(t))]^2 + [Im(s_a(t))]^2}, \quad (3)$$

$$\phi_{inst}(t) = arctan\frac{Im(s_a(t))}{Re(s_a(t))} + \pi \times sgn(Im(s_a(t))). \quad (4)$$

The instantaneous amplitude is first evaluated to derive the time-of-flight (TOF) of the front-wall echo (FWE) and back-wall echo (BWE) (see Figure 1a). The instantaneous phase associated with the FWE is set to ϕ_0. The instantaneous phase at the fundamental ply-resonance is used for basic interply tracking [7]. Practically, the interpolated $\phi_0 - \pi/2$ position in each 2π phase cycle is tracked, and the corresponding TOF indicates the depth positions of the interplies (see Figure 1c). Though, the dominating nature of the FWE and BWE makes the extraction of nearby interplies problematic [5, 7]. In order to reduce this dominating effect, the instantaneous phase at the 2nd-harmonic ply-resonance is analyzed to derive these nearby interplies (see Figure 1b). Practically, the interpolated $\phi_0 - \pi/2$ positions in the second and the second-to-last 2π phase cycles are tracked, and the corresponding TOF indicate the depth positions of the first and the last interplies.

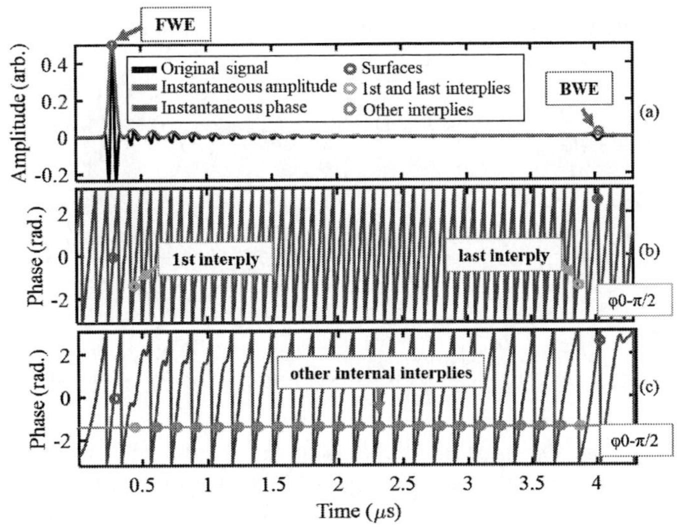

Figure 1 Illustration of the analytic-signal analysis on a simulated A-scan signal: (a) Extraction of the front wall and back wall echoes by the instantaneous amplitude of the original signal, (b) extraction of the first and the last interplies using the instantaneous phase of the filtered signal at the 2nd-harmonic ply-resonance, and (c) extraction of other interplies using the instantaneous phase of the filtered signal at the fundamental ply-resonance.

2.2 Structure tensor process to extract the out-of-plane ply orientation

The structure tensor process is a flexible and robust method for determining structure orientations in a smoothly varying field [8]. In this study, the structure tensor process is applied to the instantaneous phase of the analytic-signal. For the function $\phi_{inst}(x, y, t)$, the structure tensor is defined as [9]

$$S_G(p) = \int G_\tau(r) S_0(p-r) dr, \tag{5}$$

with $p = (x, y, t)$ and $r \in R^3$. $G_\tau(r)$ is a 3D Gaussian smoothing kernel according to

$$G_\tau(r) = \frac{1}{(\sqrt{2\pi}\tau)^3} exp(-\frac{|r|^2}{2\tau^2}), \tag{6}$$

with τ the standard deviation. $S_0(p)$ corresponds to

$$S_0(p) = \begin{bmatrix} (\phi_x(p))^2 & \phi_x(p)\phi_y(p) & \phi_x(p)\phi_t(p) \\ \phi_x(p)\phi_y(p) & (\phi_y(p))^2 & \phi_y(p)\phi_t(p) \\ \phi_x(p)\phi_t(p) & \phi_y(p)\phi_t(p) & (\phi_t(p))^2 \end{bmatrix}, \tag{7}$$

with $\phi_x(p)$, $\phi_y(p)$, $\phi_t(p)$ representing the three partial derivatives of $\phi_{inst}(\mathrm{x},\mathrm{y},t)$. The partial derivatives are obtained as follows [4]

$$\phi_x(p) = \cos\phi_{inst}(\mathrm{p})\frac{\partial\sin\phi_{inst}(\mathrm{p})}{\partial x} - \sin\phi_{inst}(\mathrm{p})\frac{\partial\cos\phi_{inst}(\mathrm{p})}{\partial x}, \quad (8)$$

$$\phi_y(p) = \cos\phi_{inst}(\mathrm{p})\frac{\partial\sin\phi_{inst}(\mathrm{p})}{\partial \mathrm{y}} - \sin\phi_{inst}(\mathrm{p})\frac{\partial\cos\phi_{inst}(\mathrm{p})}{\partial \mathrm{y}}, \quad (9)$$

$$\phi_t(p) = \cos\phi_{inst}(\mathrm{p})\frac{\partial\sin\phi_{inst}(\mathrm{p})}{\partial \mathrm{t}} - \sin\phi_{inst}(\mathrm{p})\frac{\partial\cos\phi_{inst}(\mathrm{p})}{\partial \mathrm{t}}, \quad (10)$$

where the central-difference kernels are applied to the trigonometric function values of the instantaneous phase. The gradients of the unwrapped instantaneous phase are obtained according to Eqs (8), (9), and (10). An additional 3D Gaussian kernel smoothes the trigonometric function values of the instantaneous phase G_σ before the derivation of the phase gradients. The kernels G_τ and G_σ have standard deviations of τ, and σ respectively. Finally, the eigenvectors $\mathbf{v}$ and eigenvalues λ_v of the tensor $S_G(p)$ are obtained which hold information about the structure's orientation at point p:

$$S_G(p)\mathbf{v} = \lambda_v\mathbf{v}, \quad (11)$$

From this, the local out-of-plane ply angle can be extracted by evaluating the principal eigenvector, as shown in Figure 2.

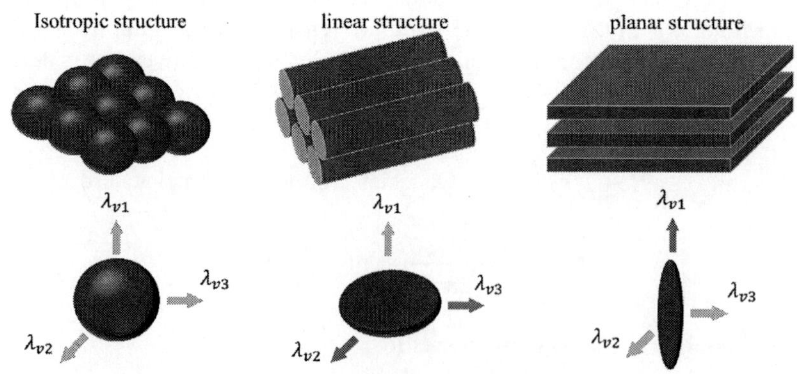

Figure 2 Illustration of extracting the principal direction of planar structures in volumetric dataset using structure tensor process.

2.3 Gabor filter-based Information Diagram to extract in-plane fber direction

The GF-ID automatically constructs a spatial 2D Gabor filter with optimal orientation and wavelength to estimate the local fiber direction. A 2D Gabor filter $h(x, y)$ with arbitrary orientation θ is given by [10]

$$h(x,y) = \exp\left[-\left(\frac{x'^2}{2\sigma_x^2}+\frac{y'^2}{2\sigma_y^2}\right)\right] \times \exp\left[j\left(2\pi\frac{x'}{\lambda_s}+\phi_s\right)\right], \quad (12)$$

$$x' = x\cos\theta - y\sin\theta, \quad (13)$$

$$y' = x\sin\theta + y\cos\theta, \quad (14)$$

where x and y are the spatial coordinates, λ_s and ϕ_s are the wavelength and phase of the sinusoidal plane wave. The parameters σ_x and σ_y determine the space constants of the Gaussian envelope along the x- and y-axes, respectively:

$$\sigma_x = \frac{\lambda_s}{\pi}\sqrt{\frac{ln\,2}{2}} \times \frac{2^b+1}{2^b-1}, \quad (15)$$

$$\sigma_y = \frac{\lambda_s}{\gamma\pi}\sqrt{\frac{ln\,2}{2}} \times \frac{2^b+1}{2^b-1}, \quad (16)$$

where b and γ are the spatial frequency bandwidth and the spatial aspect ratio, respectively, controlling the kernel shape. These parameters could be optimized to fit particular features better. In this chapter, however, they are fixed and have the values b=1 and γ=0.7.

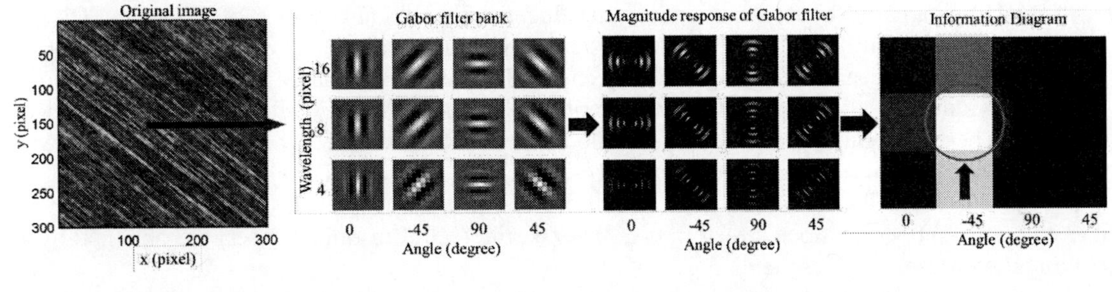

Figure 3 Schematic of the procedure of the GF-ID method: (a) the original image $I(x, y)$, (b) the Gabor filter bank with various wavelengths (horizontal) and orientations (vertical), (c) the magnitudes of the Gabor responses of the original image, and (d) the ID of the certain pixel.

A 2D Gabor filter is applied on an image $I(x,y)$ (see Figure 3). The normalized Gabor response $r^n_{\theta,\lambda_s}(x,y)$ is then obtained as follows [11]:

$$r^n_{\theta,\lambda_s}(x,y) = \frac{1}{2\pi\sigma_x\sigma_y} h(x,y) * I(x,y), \qquad (17)$$

where $*$ denotes the convolution operator. A 2D Gabor filter bank with various wavelengths and orientations is applied to $I(x,y)$. Considering that the Information Diagram method becomes unstable when the noise features fit one of the wavelengths, an additional 2D Gaussian kernel could be applied to $r^n_{\theta_0,\lambda}(x,y)$:

$$G(x,y) = \frac{2}{\pi K^2 {\lambda_s}^2} \exp\left(-\frac{2(x^2+y^2)}{K^2 {\lambda_s}^2}\right), \qquad (18)$$

where K governs the smoothness of the kernel. The K-value is selected as 1 in order to balance the sensitivity to local variations and noise. Application of the filter kernel $G(x,y)$ to $r^n_{\theta,\lambda_s}(x,y)$ finally yields the filtered normalized Gabor response $\tilde{r}^n_{\theta,\lambda_s}(x,y)$:

$$\tilde{r}^n_{\theta,\lambda_s}(x,y) = G(x,y) * r^n_{\theta,\lambda_s}(x,y), \qquad (19)$$

The magnitude of the filtered normalized Gabor response $\tilde{r}^n_{\theta,\lambda_s}(x,y)$ as a function of the 2D Gabor filter orientation and wavelength parameters forms the $ID_{x,y}(\theta,\lambda_s)$ (see Figure 3). The maximum in the $ID_{x,y}(\theta,\lambda_s)$ is adopted to determine the local fiber orientation at a specific pixel (x,y) [12]. Applying this procedure for each pixel in the original image then represents the local in-plane fiber orientation.

3. Experimental setup

A 15 MHz spherically focused immersion transducer with a focal distance of 25.4 mm (Olympus V313) and an element diameter of 6.35 mm is employed. The transducer works in the PE mode for raster scanning. This transducer is specifically chosen because it provides sufficient lateral resolution, high bandwidth, and a good signal-to-noise ratio (SNR) through the whole depth for the inspected laminate. The focal point was put in the middle of the CFRP laminate, which is immersed in water. The ultrasonic response is recorded using a 14-bit acquisition card (PXIe-5172) at a sampling frequency of 250 MS/s. The scanning steps in both x and y directions are 0.25 mm. The scanning velocities in both x and y directions are 20 mm/s. 3D experimental ultrasonic datasets are obtained by employing the ultrasonic experiment.

The CFRP laminate is evaluated in the ultrasonic experiments. The CFRP laminates have a stacking sequence of $[45/0/-45/90]_{3S}$ with 24 plies and a total thickness of approximately 5.5 mm. The laminates are produced in an autoclave. The fiber volume fraction of the laminates is approximately 60%.

4. Experimental results

The pU-CT methodology, as introduced in Section 2, is applied to the experimental ultrasonic datasets. In order to select the fundamental and 2nd-harmonic ply-resonance frequency ranges for hybrid analytic-signal analysis in the 1st step, the log-Gabor filters are applied to the time-domain signals. The instantaneous phase at the 2nd-harmonic ply-resonance is used for the structure tensor process in the 2nd step. Moreover, the GF-ID method does not require the fundamental or the 2nd harmonic ply-resonance. Thus, a log-Gabor filter around the center frequency of the transducer is applied for noise filtering in the 3rd step. The parameters of the log-Gabor filters are shown in Table 1. The center frequency f_0 is determined based on the ply-resonance, the σ_0 is chosen according to the previous study [5].

Table 1 The parameters of the applied log-Gabor filters

Function	f_0 (MHz)	σ_0	-6 dB bandwidth (MHz)
Fundamental ply-resonance	6.3	0.7	3.94 - 9.13
2nd-harmonic ply-resonance	12.8	0.8	9.61 - 16.26
Noise filtering	15	0.7	9.86 - 22.83

The structural interfaces, out-of-ply orientations α, and in-plane fiber direction θ are extracted for the healthy and the impacted CFRP laminates. The reconstructed ply thickness map, out-of-plane ply orientation map, and in-plane fiber direction map for the healthy CFRP laminate are displayed in Figure 4. The slices of the ply thickness, out-of-plane ply orientation, and in-plane fiber direction at 7th, 14th, and 21st plies for the CFRP laminate are displayed in Figure 5. The slices are aligned with the ply thickness map in Figure 4a.

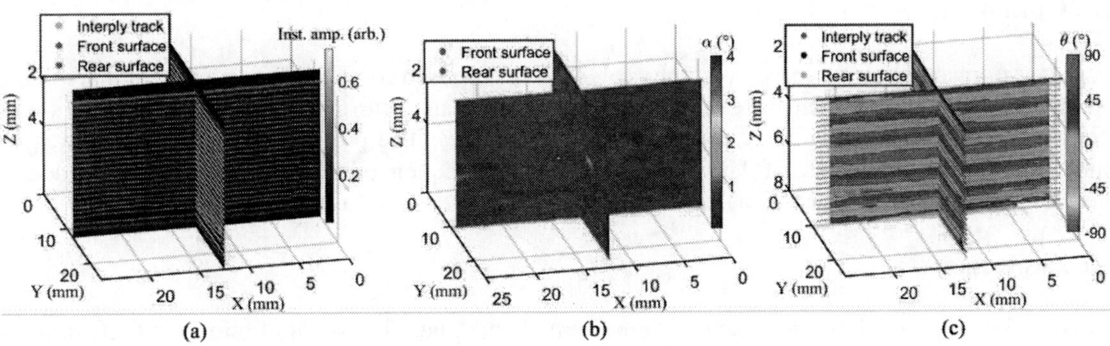

Figure 4 (a) Ply thickness map, (b) out-of-plane ply orientation map, and (c) in-plane fiber direction map for the CFRP laminate.

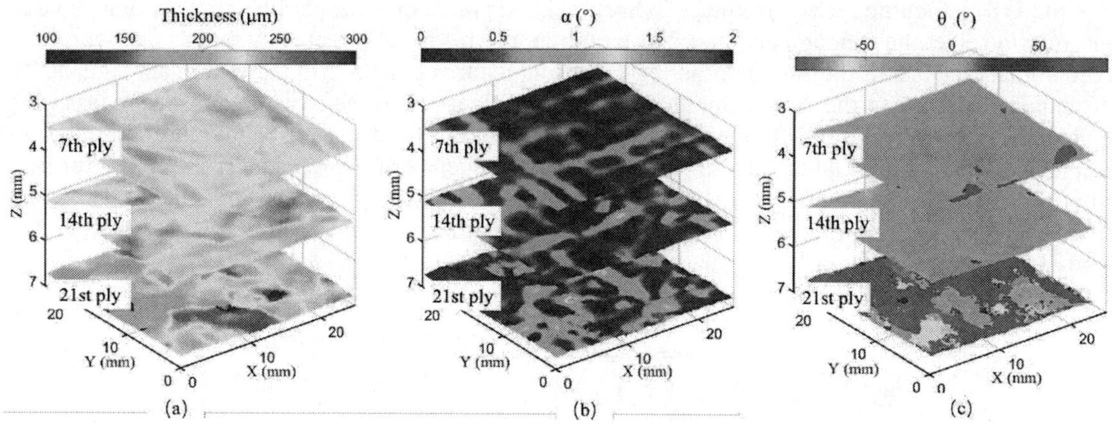

Figure 5 Ply-aligned slices of (a) Ply thickness, (b) out-of-plane ply orientation, and (c) in-plane fiber direction at 7th, 14th, and 21st plies for the CFRP laminate.

The information of the interply tracks within the materials is provided. The out-of-plane ply orientation maps provide information about the local plastic deformation inside the bulk of the CFRP laminate. Finally, the global stacking sequence and local in-plane fiber directions are extracted. However, the maps in deep layers show many deviations due to the significant ultrasonic attenuation and low SNR in deep plies.

5. Conclusion

A planar ultrasound computed tomography (pU-CT) is proposed for the reconstruction of the volumetric fiber architecture of multi-layer composites. It employs an ultrasonic pulse-echo scanning method coupled to different tomographic reconstruction approaches. The proposed pU-CT method was demonstrated on a 24-layer CFRP laminate with a stacking sequence $[45/0/-45/90]_{3S}$, and good reconstruction results have been obtained.

Acknowledgments

This work is supported by China Scholarship Council and Ghent University Preference CSC program (No. 01SC0219), Bijzonder OnderzoeksFonds Ghent University (BOF – 01N01719), and the National Natural Science Foundation of China (No. 52035013).

References

1. C. Meola, et al., 'Impact Damaging of Composites Through Online Monitoring and Non-destructive Evaluation with Infrared Thermography', Ndt& E Int, vol. 85, pp. 34-42, January 2017.
2. S. Carmignato, et al., 'Industrial X-ray Computed Tomography', Springer, 2018.

3. L. Nelson and R. Smith, 'Fibre Direction and Stacking Sequence Measurement in Carbon Fibre Composites using Radon Transforms of Ultrasonic Data', Composites Part A: Applied Science and Manufacturing, vol. 118, pp. 1-8, March, 2019.
4. L. Nelson, et al., 'Ply-orientation Measurements in Composites using Structure-tensor Analysis of Volumetric Ultrasonic Data', Composites Part A: Applied Science and Manufacturing, vol. 104, pp. 108-119, January 2018.
5. X. Yang, et al., 'Parametric Study on Interply Tracking in Multilayer Composites by Analytic-signal Technology,'' Ultrasonics, vol. 111, pp. 106315, March 2021.
6. X.Y. Yang, et al., 'Comparative Study of Ultrasonic Techniques for Reconstructing the Multilayer Structure of Composites', Ndt& E Int, vol. 121, pp. 102460, July 2021.
7. R.A. Smith, et al., 'Ultrasonic Analytic-signal Responses from Polymer-matrix Composite Laminates', IEEE transactions on ultrasonics, ferroelectrics, and frequency control, vol. 65, no. 2, pp. 231-243, February 2017.
8. P. Pinter, et al., 'Comparison and Error Estimation of 3D Fibre Orientation Analysis of Computed Tomography Image Data for Fibre Reinforced Composites', Ndt& E Int, vol. 95, pp. 26-35, April 2018.
9. A.R. Khan, et al., '3D Structure Tensor Analysis of Light Microscopy Data for Validating Diffusion MRI', Neuroimage, vol. 111, pp. 192-203, May 2015.
10. P. Kruizinga and N. Petkov, 'Nonlinear Operator for Oriented Texture', IEEE Transactions on image processing, vol. 8, no. 10, pp. 1395-1407, October 1999.
11. J. Kamarainen, et al., 'Fundamental Frequency Gabor Filters for Object Recognition', Proc. Object recognition supported by user interaction for service robots, IEEE, pp. 628-631, August 2002.
12. P. Moreno, et al., 'Gabor Parameter Selection for Local Feature Detection', Proc. Iberian conference on pattern recognition and image analysis, Springer, pp. 11-19, 2005.
13. A. Standard, 'D7136/D7136M-15, Standard Test Method for Measuring the Damage Resistance of a Fiber-Reinforced Polymer Matrix Composite to a Drop-Weight Impact Event', ASTM International, West Conshohocken, PA, 2015.
14. S. Spronk, et al., 'Comparing Damage from Low-velocity Impact and Quasi-static Indentation in Automotive Carbon/epoxy and Glass/polyamide-6 Laminates', Polymer Testing, vol. 65, pp. 231-241, February 2018.

Convergence Study on Wave Propagation in a Concrete Beam

Alaa DIAB[1], Tamara NESTOROVIĆ [2]

1 Mechanics of Adaptive Systems, Ruhr University Bochum, Germany alaa.diab@rub.de

2 Mechanics of Adaptive Systems, Ruhr University Bochum, Germany tamara.nestorovic@rub.de

Abstract:

This paper will show a convergence study on wave propagation in a concrete beam. As the wave propagation analysis is extremely dependent on two important parameters in the simulation to get correct results, many variations of both parameters will be observed. The first parameter is the element size used in the model, and the second parameter is the time increment of the analysis steps. This whole study is implemented as a part of a larger project that will study the possibility of detecting damages in different structures in general and in massive structures in specific. Therefore, in the second part of this paper, another model of the concrete beam with a crack will be also implemented. So, this paper will be used as clarification of the numerical model and as a basis for values decision of the two critical parameters should be taken with less memory consumption and time cost. The most challenging difficulty that appeared during this study, is how to decide on the values of element-size without affecting the critical conditions of modeling a wave using explicit analysis in FEM. The wave used for excitation in the analyses was a cycled Hanning wavelet applied at one side of the beam, and the response was detected at the other side of the beam. During the analysis, a strange effect of the beam response was observed, where the response at a specific element size is shifted a bit in comparison to the response at another element size attempt. Furthermore, the peak magnitude of the response is also slightly changing through different element size attempts. For a given beam geometry and material properties, the best value of element size can be acquired after a number of iterations. Trying a smaller time increment with the same mesh size has no significant effect on the response as long as the values used are not exceeding the critical value defined by the Courant–Friedrichs–Lewy condition.

Keywords: Finite elements method (FEM), Convergence study, Wave propagation, Explicit analysis, Crack detection

1. Introduction

Most of the structures nowadays are depending mainly on concrete as building materials. This is due to its properties that satisfies both the serviceability and ultimate international design limits. This have been used in variety of structures such as skyscrapers [1], huge bridges [2], parking buildings, airports, hospitals, dams, etc. Even though, concrete has a relatively satisfying compression strength, its tensile strength is very small compared to its compression strength. Due to this large difference between its behavior in both situations, reinforcement is always needed in the places where a concrete element is expected to work against tensile forces. However, this does not make it invulnerable. As a result, cracks occur if tension exceeds the concrete tensile strength and lead the reinforcement steel to be exposed to air and natural environment if it exists. On one hand, if the reinforcement is exposed to hostile environment, it could be damaged over the time through several phenomena such as corrosion, creep, etc. On the other hand, existing of cracks in a structural element changes the functionality of this element and reduces its resistance in the neighborhood of cracks. Furthermore, appearing of cracks and their subsequent growth are not only depending on the tensile strength being exceeded, but also on many other reasons. If all these reasons are combined together, it would be very complicated to detect the main cause of cracks or avoid their spread in concrete structures. Some of these reasons are location, length, width, shape, age of the crack, and also the number of cracks taking place in the structure.

Several methods were already implemented to monitor concrete systems and are categorized into two main types. Destructive evaluation methods and non-destructive evaluation (NDE) or structural health monitoring (SHM). As the destructive methods results in damaging the structure, it is not so much common to be used nowadays. Nonetheless, many non-destructive methods emerged in the last decade such as acoustic emission testing (AE), electromagnetic testing (ET), laser testing methods (LM), microwave testing, ultrasonic testing (UT) [3] . These methods have already approved its performance and they give a great contribution to the structural monitoring and testing field. Many of the NDE methods are relatively expensive and demand special tools. However, the research interest is concentrating on developing a comparatively cheap method that presents a convenient detection of structural damage if exists. This method is based on using piezoelectric transducers (PZT) as actuators and sensors recoding a wave propagating through the structure. After that, the wave record will be used to calculate the damage index which will give the notification of existing damage if its value is changed [4].

In this paper, the planned experiment is simulated using FEM to clarify its applicability. However, due to the sensitivity of this wave propagation simulation, the aim is to tune the modeling parameters in such a way that the simulation cost regarding elements size and the time increments is reduced without influencing the accuracy of the captured response.

2. Methodology:

In this paper, an approximation of the PZT is implemented based on the experiment done in [5] where a linear relationship between the wave frequency and the response of the concrete beam is found. This gives the advantage of reducing the analysis cost for the model. Thus, the PZT is substituted by a concentrated force applied in the location of the actuator PZT and the sensor PZT is modeled as an observation point which will collect the response for each time step.

2.1. Analysis of wave propagation

For the case of modeling a concrete beam, the wave propagation can be provided through a PZT actuator. Every wave is mainly composed of two types; bulk and surface waves. The bulk waves are generally propagated through the interior of the beam as longitudinal waves (L) and shear waves (S). On the other hand, surface waves are propagating primary along the beam surface or between two surfaces of the concrete element.

The concrete material defined for this paper is considered as an isotropic elastic material. Additionally, the lateral dimensions of the concrete beam are big enough compared to the wavelength to be considered. Based on these considerations, the equations relating the behavior parameters and the elastic waves velocities can be described as followed [6]:

$$c_L = \sqrt{\frac{\lambda + 2\mu}{\rho}} , \quad c_S = \sqrt{\frac{\mu}{\rho}} \tag{1}$$

Where the Lame constants (μ, λ) required for the calculation can be acquired depending on the material properties as followed:

$$\mu = \frac{E}{2(1+\upsilon)} , \quad \lambda = \frac{E\,\upsilon}{[(1-2\upsilon)(1+\upsilon)]} \tag{2}$$

where

c_L, c_S - the velocities of the longitudinal and shear waves.
E, υ, ρ - Young's modulus, Poisson's ratio and material density respectively.

2.2. *Wave propagation modeling*

For the purpose of wave propagation modeling, the Explicit Finite Element Method (EFEM) is used. The method is derived basically from the second Newton's law [7]:

$$F = M\,.\,A \tag{3}$$

where F, M, A are the external forces, mass, acceleration matrices respectively. The previous equation represents a system of second order differential equations in time. As the explicit algorithm is going to be used, the method of central differences is applied to find the approximate values of acceleration, velocity and displacement.

The terms of Eq. (3) can be calculated as follows:

$$\begin{aligned} F_i^M &= -\int_\Omega (\sigma_{ij}\Phi_{M,j})d\Omega + \int_\Omega f_i\Phi_M\, d\Omega + \int_{\Gamma_s} g_i\Phi_M\, d\Gamma \\ M_{MN} &= \int_\Omega (\rho\, \Phi_M\, \Phi_N)\, d\Omega \\ A_i^N &= \ddot{u}_i^N(t) \end{aligned} \tag{4}$$

Further description of Eq. (4) terms can be found in [7]. To present a brief summary on the central difference method, the time range is assumed to be within the range $(0, T)$ which is considered to be divided regularly into N even sub-domains $[t_n, t_{n+1}]$, where $0 = t_0 < t_1 < \cdots < t_N = T,\ \ t_{n+1} - t_n = \Delta t = T/N$.

The following step includes approximating the time derivatives of both velocity and acceleration using the finite difference method as represented in the following in vector form:

$$\begin{aligned} \partial_t\, \boldsymbol{u}_{n+1/2}^h &= \frac{(\boldsymbol{u}_{n+1}^h - \boldsymbol{u}_n^h)}{\Delta t} \sim \dot{\boldsymbol{u}}_{n+1/2}^h \\ \partial_{t^2}\, \boldsymbol{u}_n^h &= \frac{(\partial_t\, \boldsymbol{u}_{n+1/2}^h - \partial_t\, \boldsymbol{u}_{n-1/2}^h)}{\Delta t} = \frac{(\boldsymbol{u}_{n+1}^h - 2\,\boldsymbol{u}_n^h + \boldsymbol{u}_{n-1}^h)}{\Delta t^2} \\ &\sim \ddot{\boldsymbol{u}}_n^h \sim \frac{(\dot{\boldsymbol{u}}_{n+1/2}^h - \dot{\boldsymbol{u}}_{n-1/2}^h)}{\Delta t} \end{aligned} \tag{5}$$

$$\begin{aligned} \partial_t\, \boldsymbol{u}_{n+1/2}^h &= \partial_t\, \boldsymbol{u}_{n-1/2}^h + \partial_{t^2}\, \boldsymbol{u}_n^h\, \Delta t \\ \boldsymbol{u}_{n+1}^h &= \boldsymbol{u}_n^h + \partial_t\, \boldsymbol{u}_{n+1/2}^h\, \Delta t \end{aligned} \tag{6}$$

where $\boldsymbol{u}$ represents the general displacements.

Substituting the acceleration in Eq. (3) by its central difference approximation in Eq. (5,6):

$$\partial_{t^2}\, \boldsymbol{u}_n^h = \boldsymbol{M}^{-1}\, \boldsymbol{F}_n \tag{7}$$

One of the greatest pros for using the explicit finite element method is the diagonal mass matrix generated for the model, this provides extreme efficiency compared to the implicit method [8]. The advantage appears during the step of calculating the acceleration through the second Newton's law Eq. (7), where the inverse of the mass matrix can be easily calculated.

2.3. *Explanation of used technique*

In the following sections, a description of the critical parameters for implementing a FEM model for the purpose of wave propagation analysis is presented. Furthermore, a suitable technique is presented to guarantee that none of the critical limits is reached during the convergence study.

2.3.1. *Critical considerations*

As the analyzing method used for this research is explicit, it is important to mention that the time increment can be crucial, for example, considering a specific value of time increment, relatively greater error is arising for the explicit method in comparison to the implicit method. Nonetheless, we can guarantee a rewarding accurate solution using the explicit method by choosing adequately small time increments (Δt). However, the time increment is not allowed to have values greater than a certain threshold that depends on the Peclet number and on the smallest mesh size in the model [8]. The governing rule for determining the critical time increments is given as followed:

$$\Delta t_{crit,1} = \left(\frac{L_e}{c_d}\right) \tag{8}$$

where:

L_e	the smallest characteristic dimension of the finite element
c_d	the wave velocity (the maximum wave velocity).

From Eq. (8), we can see that for fine mesh, small time increment is demanded. This has to be calculated for each time step. Thus, when smaller time increment is applied, the computational cost of analysis becomes very high.

Another determining rule for choosing the time increment depends on the highest frequency (ω_{max}) obtained for the model with the smallest finite element size in non-damping systems [9]:

$$\Delta t_{crit,2} = \frac{1}{20 * \omega_{max}} \tag{9}$$

As $\Delta t_{crit,1}$ and $\Delta t_{crit,2}$ are the greatest allowed value for the time increment, the final governing equation to determine the time increment values is taken as the smaller value of both limits:

$$\Delta t \le \min(\Delta t_{crit,1}, \Delta t_{crit,2}). \tag{10}$$

Another important factor for the success of wave propagation simulation is the number of finite elements per wavelength, where it was recommended in [10] to use a greater or equal to seven elements per wavelength.

2.3.2. *Choosing the experimental values*

For the convergence study, the parameter of interest is to be varied under the previously mentioned conditions. In the first part of the research, the meshing size is going to be varied. However, the change in its value will also change the critical time increment in Eq. (8). Thus, the time increment will be assigned to a fixed value that satisfies the threshold corresponding to each testing value of the mesh size.

As the number of elements per wavelength is to be considered in wave propagation analysis, the wavelength must be subsequently calculated for each attempt value of the element size. This can be acquired through the following equation:

$$wave\ length = \frac{v_{min}}{\omega_{max}} \tag{11}$$

where v_{min} is the minimum velocity of the implemented wave and ω_{max} is the maximum frequency of the wave which will be considered to be 40% greater than the signal frequency.

For the second part of the research, a stable value of mesh size will be taken as a fixed value and the time increment is to be varied. This will give a general view of the influence of the time increment change on the response of the concrete beam under wave propagation excitation.

It is important to mention, that for both parts of the research, the response captured at the sensor location is to be used as judging factor. Further information about the models implemented is presented in the following paragraph.

3. Numerical Simulation and Results

3.1. Defining the parameters

3.1.1. The material properties

A concrete beam is modeled as a linear elastic material with the dimension of $0.4 \times 0.127 \times 0.127$ [m] with two approximated PZT at points A and S (Figure 1). The material properties of the concrete used for modeling are the same for both parts of the research and they are shown in Table 1.

Table 1. Material properties of concrete

Young's modulus [GPa]	30
Poisson's ratio	0.2
Density [kg/m^3]	2400

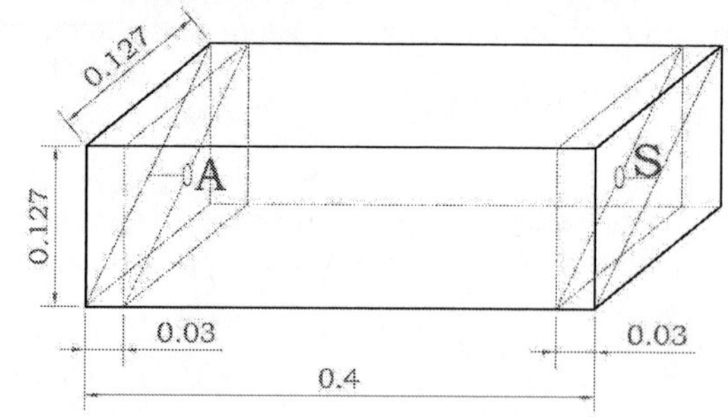

Figure 1. Geometry of the model [m]

This model will be used for both parts of the research. The main difference between the two parts is that the first part will be oriented to implement the convergence study by changing the values of element size in the model and fixing a suitable value of the time increment, while the second part of the research will be concentrated on finding the effect of changing the time increment for a fixed value of element size.

For both parts of the research, the same function of the displacement change is defined as a 3.5-cycle Hanning-windowed tone burst (Figure 2), which can be described by

$$P_{(t)} = \begin{cases} \left[1 - cos\left(\frac{2\pi}{N}t\right)\right] sin(2\pi f t), & \text{for} \quad 0 \leq t \leq \frac{N}{f} \\ 0\,, & \text{for} \quad t > \frac{N}{f} \end{cases} \tag{12}$$

where

f	signal frequency (in the model 100 kHz)
N	number of cycles (in the model 3.5 cycles)
t	time (duration of the signal) (in the model 3.5×10^{-5} s)
$P_{(t)}$	the output signal.

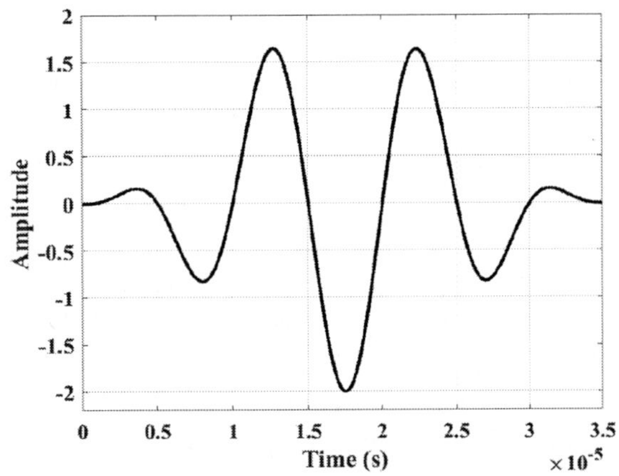

Figure 2. Excitation signal of 3.5-cycle Hanning-windowed tone burst

The analysis is done using the ABAQUS/EXPLICIT package by applying C3D8R eight-node prismatic finite elements with reduced integration and Hourglass control.

3.1.2. Selection of the wave propagation parameters

For the purpose of implementing the convergence study, the wave length is calculated according to the predefined wave characteristics through Eq. (11). Accordingly, the approximated number of elements for each attempt is determined as the division of the wave length by each mesh size.

As the number of elements to be considered in this study is recommended to be more or equal to seven, we are going to try to start the mesh size values variation that gives ten elements per wave length. Table 2 shows the possible values to consider for the convergence study.

Table 2. Possible characteristic values for the convergence study

Mesh size [m] x10^{-3}	1.55	1.50	1.45	1.40	1.35	1.30
Nr. of elements	10.5	10.9	11.2	11.6	12.1	12.5
$\Delta t_{crit,1}$ [s] x10^{-7}	4.16	4.02	3.89	3.76	3.62	3.49
$\Delta t_{crit,2}$ [s] x10^{-7}	3.57					
Mesh size [m]x10^{-3}	1.25	1.20	1.15	1.10	1.05	1.00
Nr. of elements	13.0	13.6	14.2	14.8	15.5	16.3
$\Delta t_{crit,1}$ [s] x10^{-7}	3.35	3.22	3.09	2.95	2.82	2.68
$\Delta t_{crit,2}$ [s] x10^{-7}	3.57					

As the wave propagation analysis is very sensitive to the element length and time increment used for the explicit analysis, the following values in Table 3 shows the values of element length or mesh size being considered for each attempt of the analysis for the first part of the research. It is important to mention that the time increment is considered to be equal to a fixed value that satisfies the conditions in Eq. (10).

Table 3. Element size values for the convergence study

Mesh size [m] x10^{-3}	1.30	1.25	1.20	1.15	1.10	1.05	1.00
Time increment [s]	2.10 x10^{-7}						
Model number [-]	1	2	3	4	5	6	7

For the second part of the research, the value of time increment is assigned to two different values that satisfy Eq. (10) and the condition of number of elements per wave length, where the mesh size is fixed to the value ($1.05 \cdot 10^{-3}$ m) which was found as the stable value in the convergence study.

The third part of the research includes a concrete beam with the same characteristics with one crack in the middle of the beam from the upper side with a crack length of 0.03175 m which is 25% of the beam depth. This will present the basic step in the procedure of crack detection through recording the wave propagation response at the sensor position.

3.2. Simulation models

Both models implemented in this research are represented in Figure 3. The one on the left, is for an intact concrete beam which is used for the convergence study and the time increment variation example. While the second model on the left, presents a basic mechanism of detecting a crack using wave propagation observation.

Figure 3. Concrete beam models; left: intact beam, right: damaged beam

Additionally, the simulation response at four selected time steps for the damaged model is presented in Figure 4. There we can see the reflected waves from the crack surface and the arrival of part of the original propagated wave to the sensor.

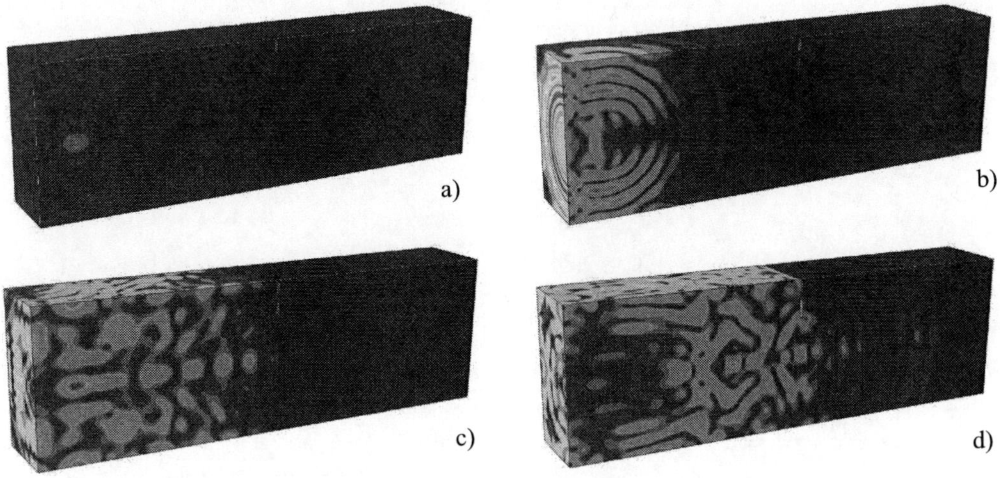

Figure 4. Wave propagation of the damaged concrete beam at different time steps:
a) $t = 2.52 \cdot 10^{-5}$ s, b) $t = 3.40 \cdot 10^{-5}$s, c) $t = 7.88 \cdot 10^{-5}$ s, d) $t = 11.97 \cdot 10^{-5}$s

3.3. Responses comparison

In order to implement the convergence study, the response at the sensor is recorded for each attempt of the mesh size. Figure 5 represents the observed response at the sensor starting from the moment of the wave arrival until a specific time interval. The time after that interval will not be shown because of the fact that it includes the time of arrival of other reflected waves from the surfaces of the beam. Further analysis is carried out to show the effect of changing the time increment on the analysis for a mesh size equal to ($1.05 \cdot 10^{-3}$ m), see Figure 6. The second model of the damaged beam is analyzed and its response is shown in Figure 7.

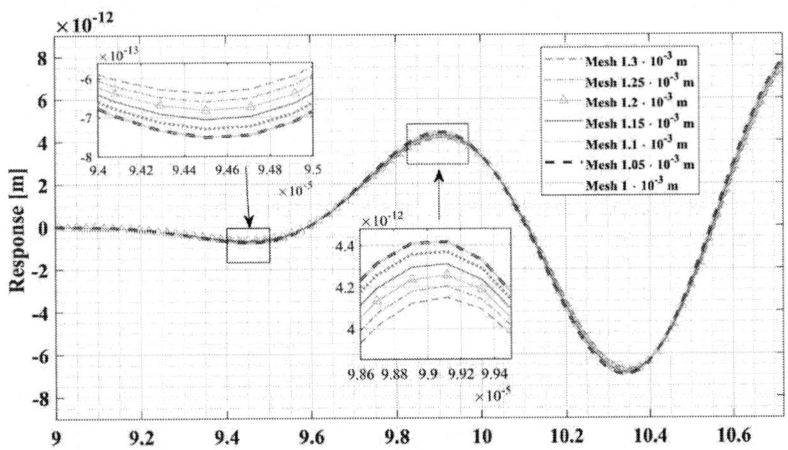

Figure 5. Responses of the beam at the sensor for variety of mesh sizes and a time increment *2.1* 10^{-7}*s*

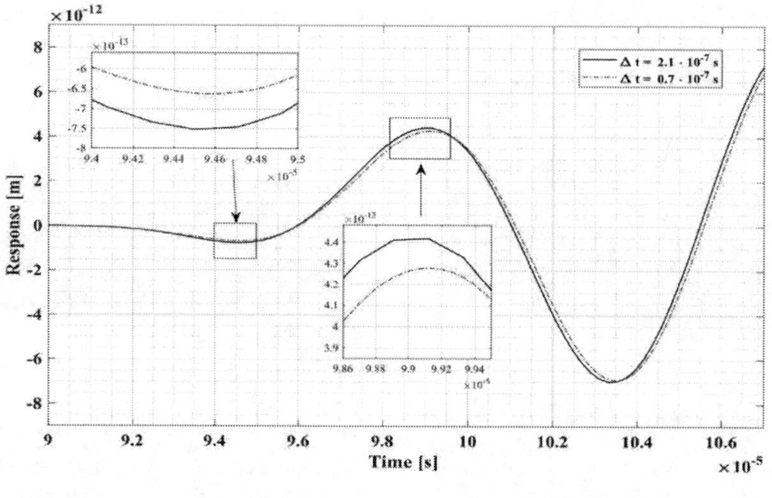

Figure 6. Responses of the beam at the sensor for variety of time increment and mesh size *1.05* 10^{-3} m

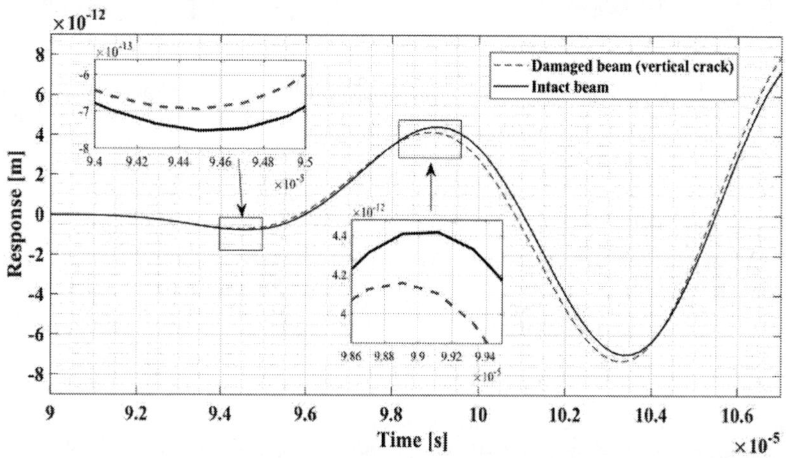

Figure 7. Responses of the beam in case intact beam (solid line) and damaged (dashed line)

4. Conclusions

As a result of the convergence study in Figure 5, after several iterations of different mesh size that the convergence point was found. For the given model, the value of $1.05 \cdot 10^{-3}$ m mesh size converges where the sixth and seventh iterations are identical. The time increment change in Figure 6 has shown that there is a slight shift of the response and a small change of the amplitude. Thus, further investigation is recommended on the tuning of this parameter after finding the suitable value of mesh size. The damaged model in Figure 7 results in a remarkable reduction of the response amplitude compared to its amplitude in the intact model. This concept will be helpful to use for further researches in non-destructive damage detection. The time of flight has not changed remarkably because the crack in the model was not opposing the direct wave path between the actuator and the sensor.

Acknowledgements:

Funded by the Deutsche Forschungsgemeinschaft (DFG, German Research Foundation) – 448696650; 77309832.

References

[1] A. Mandić Ivanković, V. Čandrlić and J. Bleiziffer, "The Largest Concrete Arch Bridge Designed of RPC 200," 2001.

[2] K. Al-Kodmany, "Skyscraper in the Twenty-First Century City: A Global Snapshot," *Buildings,* p. 175, 12 2018.

[3] twi-global.com, "What is Non-Destructive Testing (NDT)? Methods and Definition," TWI Ltd, 25 March 2019. [Online]. Available: https://www.twi-global.com/technical-knowledge/faqs/what-is-non-destructive-testing. [Accessed 13 August 2021].

[4] T. Nestorović, D. Stojić and N. Marković, "Active Structural Health Monitoring of Reinforced Concrete Structures using Piezoelectric Smart Aggregates," in *8th European Workshop On Structural Health Monitoring (EWSHM 2016)*, Bilbao, Spain, 2016.

[5] N. Marković, T. Nestorović and D. Stojić, "Numerical modeling of damage detection in concrete beams using piezoelectric patches," *Mechanics Research Communications,* vol. 64, no. 0093-6413, pp. 15-22, 2015.

[6] A. Bedford and D. Drumheller, Introduction to Elastic Wave Propagation, Chichester: John Wiley & Sons Ltd, 1994.

[7] S. Wu and L. Gu, Introduction to the Explicit Finite Element Method for Nonlinear Transient Dynamics, Hoboken, New Jersey: JOHN WILEY & SONS, INC., 2012, pp. 19-21.

[8] D. Kuzmin, A guide to numerical methods for transport equations, Erlangen-Nürnberg: Friedrich-Alexander-Universität, 2010, pp. 29-30.

[9] D. Cerniglia, A. Pantano and N. Montinaro, "3D simulations and experiments of guided wave propagation in adhesively bonded multi-layered structures," *NDT & E International,* vol. 43, no. 0963-8695, pp. 527-535, 2010.

[10] M. B. Drozdz, Efficient finite element modelling of ultrasound waves in elastic media (Ph.D. thesis), London: Imperial College, University of London, 2008.

Semi-Analytical Modelling of Perturbations in Magnetic Flux Density due to Narrow Flaws in Eddy Current NDT

Prashanth BASKARAN, Dario. J. PASADAS, Artur. L. RIBEIRO, Helena. G. RAMOS

Instituto de Telecomunicações, Instituto Superior Técnico, Universidade de Lisboa; Lisboa; Portugal; e-mail: prashanth.baskaran@tecnico.ulisboa.pt

Abstract

In this work an alternate model based on boundary element method has been proposed to evaluate the magnetic flux density components arising in eddy current nondestructive testing. The model uses the magnetic vector potential, arising from the electric dipole source, to determine the voltage induced in a hypothetical small circular loop located around the evaluation point. Based on the linear relationship between the induced voltage and magnetic flux enclosed within the hypothetical surface, the linearity constant that transforms the induced voltage to magnetic flux density can be computed. The BEM results are then corroborated with results from finite element model.

Keywords: Boundary element model, Eddy current testing, Induced voltage, Magnetic flux density, Narrow flaw.

1. Introduction

Nondestructive testing (NDT) plays a crucial role in industries to inspect materials and components for the presence of anomalies without damaging their structural integrity. Inspection on electrically conductive materials often used in aerospace, nuclear and petrochemical industries employ eddy current based NDT. Defects are generally identified by analysing the response of a magnetic field sensor or impedance of the coil or voltage induced in a secondary coil. To preview the magnetic field response due to a flaw, finite element simulations (FEM) can be used. The main advantage of FEM is that it can handle a wide geometry of problems. However, for simple geometries that are encountered in daily life such as a plate or tube, using FEM can be time consuming. In that case, if the flaw that needs to be modelled is narrow, then faster models based on boundary element method (BEM) can be utilized.

Bowler [1] developed the boundary element model for an ideal crack on the surface of a semi-infinite conductor. The argument in [1] was that, due to the presence of an ideal crack, the eddy currents are discontinuous on either side of the flaw surface and since the conductivity is the same on either side, leads to a discontinuity in the tangential component of electric field. This discontinuity is modelled by electric dipoles that are oriented normal to the flaw surface and in opposition to the incident field. The problem of an ideal crack has been, since then, extended to a flaw with narrow opening.

The distribution of the dipole density has to be computed, in order to determine the scattered electric or magnetic fields, where the scattered fields are generally represented by a Fredholm type integral equation whose kernel is the electric-electric or magnetic-electric Green's function. The dipole density, being unknown, can be computed by first discretizing the flaw surface and thus using the method of moments converting the whole system to a linear equation. This results in a discrete set of dipoles populating the flaw surface. Now, applying the vanishing incident current density to the integral equation, the dipole density can be determined.

Most of the studies on BEM in ECT application focus on determining the impedance of a coil [2] or the z component of the magnetic flux density [3]. However, for the sub-surface flaw detection, other components of the magnetic fields are also needed. Thus, focusing on rapidity in this work, we have determined the other components of the magnetic fields. The results are corroborated by the finite element model.

2. Boundary element model

The problem that we have modelled is the field scattered by a rectangular flaw, located in the sub-surface, in an aluminium plate (Al-1050). A cylindrical coil scans the flaw, positioned at a certain lift-off distance z_1. A typical eddy current scan is depicted in fig. 1. The magnetic field is evaluated at the axial centre of the coil at the point denoted in green in fig.1. The excitation field varies time harmonically as the Re$[e^{-i\omega t}]$, where ω is the angular frequency.

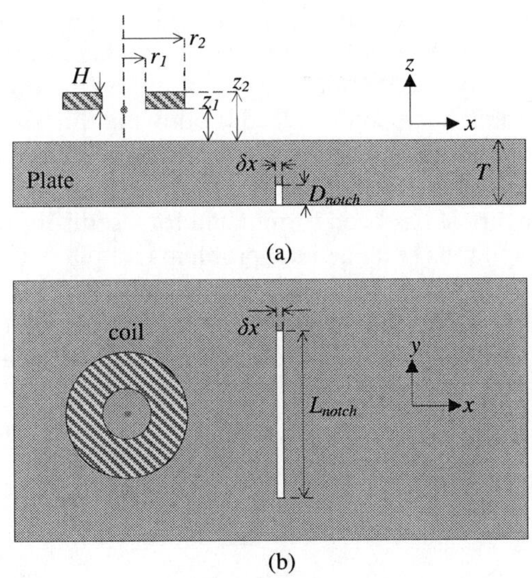

Figure 1. A typical ECT scan: (a) side view and (b) top view.

In the context of BEM formulation, the scattered electric fields due to a horizontal electric dipole source $p(r_0)$ distribution in a volume V_0 oriented along x-axis embedded in an isotropic nonmagnetic medium is usually represented as

$$E_{x_0}^{inc} = -i\omega\mu_0 \int_{V_0} dV_0 \cdot G_{xx}^{ee}(r - r_0)p(r_0) \qquad (1)$$

In (1), $E_{x_0}^{inc}$ is the unperturbed electric field that is incident normal to the flaw surface and G_{xx}^{ee} is the electric-electric Green's dyad that transforms the electric dipole source p located at r_0 to fields located at r. The equation is solved by using the method of moments and thus, it is converted to a linear system as shown in (2)

$$\boldsymbol{E}_{x_0}^{inc} = -[\boldsymbol{G}^{ee}]\mathbf{p}_{x_0} \qquad (2)$$

In eq. 2, $\boldsymbol{E}_{x_0}^{inc}$ is the incident electric field at the center of each discretized element on the flaw surface, $\boldsymbol{p}_{x_0}$ is the vector of dipole density corresponding to each cell element of the discretized flaw and $[\boldsymbol{G}^{ee}]$ is the matrix of Green's function that contains the interaction between the electric dipole source and the evaluation points.

3. Computation of the magnetic flux density

The perturbation in the magnetic flux density due to an embedded electric dipole is conventionally computed with the assistance of the magnetic-electric Green's function [3]. However, alternate models have been proposed in the literature such as [4], wherein the magnetic sensor is modelled as a small circular coil and then invoking the principle of reciprocity to determine the EMF in the hypothetical coil. Now, based on the sensitivity and the surface area of the small coil, the signal proportional to the magnetic flux density component had been deduced. Following this work, instead of determining the induced voltage through the reciprocity principle, we determined the induced voltage from the magnetic vector potential arising from the Hertz potential formulation [5]. In [5], the Hertz potential of an embedded horizontal electric dipole has been formulated for a semi-infinite plate. However, we added another boundary condition and thus extended the problem to a plate of finite thickness.

To determine the magnetic flux density components, at the desired point marked in green in fig. 1, we considered a hypothetical coil of single turn, of radius r_0 around the evaluation point as shown in fig. 2.

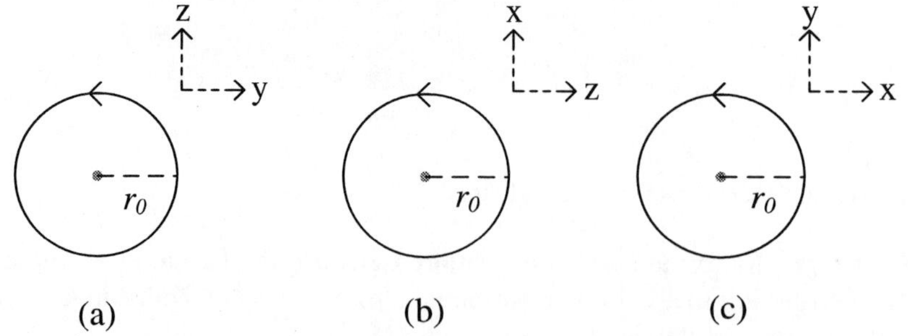

Figure 2. EMF induced in the coil whose axis is oriented along (a) x-axis, (b) y-axis, and (c) z-axis.

Now, the induced EMF for the 3 cases is given by:

$$\delta V_a = i\omega\mu_0 \oint \delta A_y \mathrm{d}y + \delta A_z \mathrm{d}z \quad (3)$$

$$\delta V_b = i\omega\mu_0 \oint \delta A_x \mathrm{d}x + \delta A_z \mathrm{d}z \quad (4)$$

$$\delta V_c = i\omega\mu_0 \oint \delta A_y \mathrm{d}y + \delta A_x \mathrm{d}x \quad (5)$$

where, δA_x, δA_y and δA_z correspond to the magnetic vector potential arising due to the embedded source, whose formulation can be arrived based on [5] for a finite thickness plate. Now, the magnetic flux through the hypothetical loop, say for the case (a), is given by,

$$\phi_a = \iint B_x \hat{x} \cdot \mathrm{d}y\mathrm{d}z\hat{x} = \iint B_x \mathrm{d}y\mathrm{d}z \quad (6)$$

Now, for a sufficiently small radius of the hypothetical loop, we can assume that the magnetic flux density component is constant within the loop and thus reducing eq. 6 to

$$\phi_a = B_x \cdot (\pi r_0^2) \quad (7)$$

Now, the induced EMF in the hypothetical coil due to the time varying magnetic flux is given by,

$$\delta V_a = i\omega\phi_a \quad (8)$$

Now, substituting eq. 4 and eq. 1 in eq. 5 and extending the same idea to case b and case c, we have finally,

$$B_x = \frac{\mu_0}{\pi r_0^2} \oint \delta A_y dy + \delta A_z dz , \quad (9)$$

$$B_y = \frac{\mu_0}{\pi r_0^2} \oint \delta A_x dx + \delta A_z dz , \quad (10)$$

$$B_z = \frac{\mu_0}{\pi r_0^2} \oint \delta A_y dy + \delta A_x dx \quad (11)$$

Now, the integrals in eq. 6 to eq. 8 can be computed by transforming them to polar co-ordinates so that the line integrals have, as a lower limit of 0 and an upper limit of 2π.

4. Results

The obtained theoretical model is corroborated by FEM based simulations. For this purpose, an excitation coil of inner radius 5 mm, external radius 10 mm and a height of 8 mm was considered. The excitation coil has 391 turns and is excited at 1 kHz. The plate used was aluminium-1050 whose electrical conductivity is $3.56 \times 10^7 \mathrm{MSm}^{-1}$ and a depth of 3 mm. A linear flaw was considered whose dimensions are 20.5 mm × 0.5 mm × 1 mm, corresponding to the length, width, and depth. Two scans were made, first along the direction of the flaw (*Oy*), and next, perpendicular to the direction of the flaw (*Ox*). The B_y and B_z components were evaluated for the first scan and B_x and B_z were evaluated for the second scan. The results for the BEM and FEM, for the case of first scan and second scan are shown in fig. 3a and fig. 3b respectively.

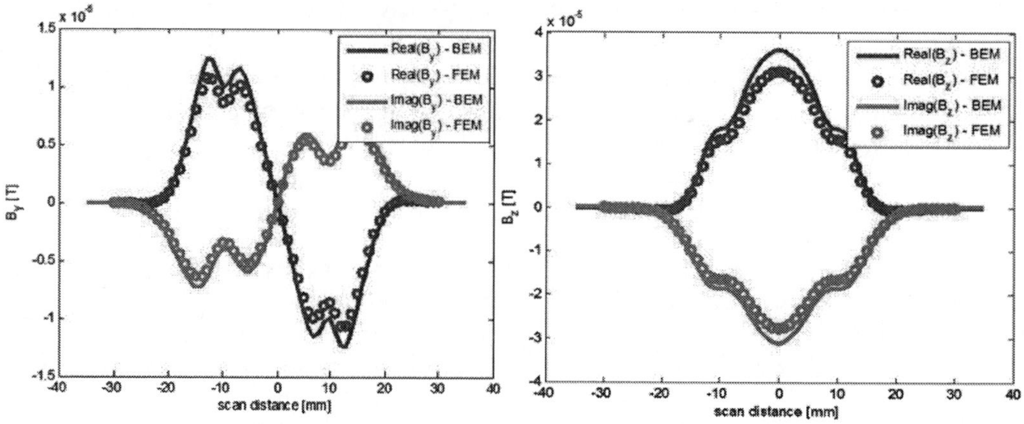

(a)

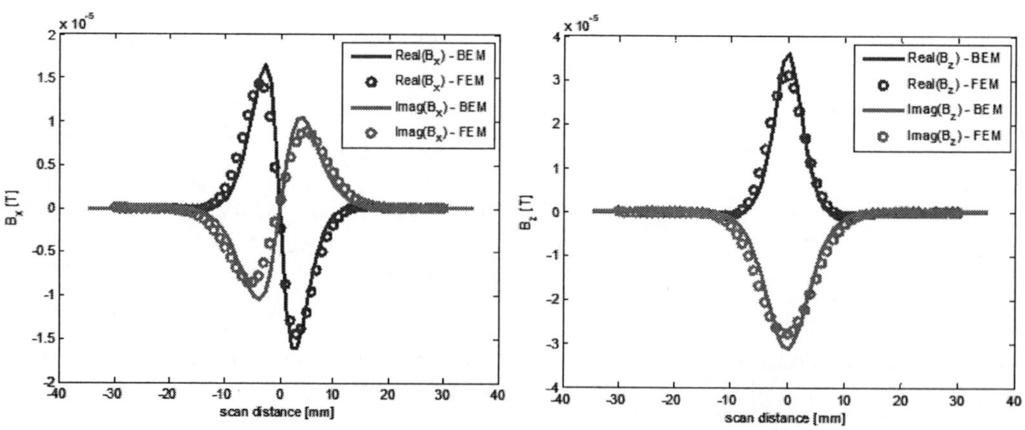

(b)

Figure 3. Comparison of magnetic flux density predictions between BEM and FEM for (a) first scan along *Oy*, and (b) second scan along *Ox*.

5. Conclusion

We have developed an alternate semi-analytical model to evaluate the perturbations in magnetic flux density components due to a narrow flaw based on BEM without the requirement of magnetic-electric Green's functions. The computation time for a single scan was approximately 10 seconds, as opposed to FEM, requiring a very high density mesh, taking nearly 15 minutes to evaluate the fields for a single position of the coil. The main advantage of this model is that it can be used to predict the magnetic field response for the case of simple geometry flaws in a rapid way and thus can be used for studies related to inverse problems and model assisted PoD of flaws, wherein a large quantity of data might be required.

References

[1] J. R. Bowler, 'Eddy-current interaction with an ideal crack. I. The forward problem,' Journal of Applied Physics, vol. 75, pp. 8128-37, 1994.
[2] R. Miorelli, C. Reboud, T. Theodoulidis, N. Poulakis and D. Lesselier, 'Efficient Modeling of ECT Sig-nals for Realistic Cracks in Layered Half-Space,' IEEE Transactions on Magnetics, vol. 49, no. 6, pp. 2886-2892, 2013.
[3] C. Voulgaraki, N. Poulakis and T. Theodoulidis, 'Theoretical Simulations and Quantitative Mag-netic Field Measurements for Eddy-Current Testing with an HTS SQUID System,' IEEE Transactions on Applied Superconductivity, vol. 23, no. 4, 2013.
[4] Roberto Miorelli, Christophe Reboud, Charitini Voulgaraki, Nikolaos Poulakis, Theodoros The-odoulidis, 'Semi-analytical solution to the 2014 eddy current benchmark problem,' AIP Conference Proceedings, 2015.
[5] Baños A., Wesley J. P., 'The Horizontal Electric Dipole in a Conducting Half-Space,' UC San Diego: Library – Scripps Digital Collection, 1953.

Performance of Time- and Frequency-Domain Analysis Techniques in Flash Thermographic Inspection of Monolithic and Sandwich Composite Parts

Gaétan POELMAN [1], Saeid HEDAYATRASA [1,2] and Mathias KERSEMANS [1]
[1] Mechanics of Materials and Structures (UGent-MMS), Department of Materials, Textiles and Chemical Engineering (MaTCh), Ghent University, Technologiepark-Zwijnaarde 46, 9052 Zwijnaarde, Belgium;
Phone: +32 9 3310434, email: Gaetan.Poelman@UGent.be, Saeid.Hedayatrasa@UGent.be, Mathias.Kersemans@UGent.be
[2] SIM Program M3 DETECT-IV, Technologiepark-Zwijnaarde 48, 9052 Zwijnaarde, Belgium;
email: Saeid.Hedayatrasa@UGent.be

Abstract
In flash thermography, the diffusive and damped characteristics of thermal waves complicate the detection of small and deep defects. To enhance the defect detectability of the recorded thermographic data, many researchers have developed a large range of advanced post-processing techniques. Most of these techniques either analyse the data in time domain or in frequency domain, however, there is no clear consensus about the optimal processing approach. In this contribution, an experimental comparison between several time- and frequency-domain analysis techniques is presented for the thermographic inspection of composite parts. In addition to laboratory coupon samples, also industrial composite components are investigated.
Keywords: Non-destructive testing (NDT), flash thermography, data processing, time domain, frequency domain, composites

1. Introduction

Due to the layered structure of composite materials, they are susceptible to internal damage features which could lead to catastrophic failure if they remain unnoticed. Therefore, a lot of effort has been dedicated over the past decades towards the development of reliable non-destructive testing (NDT) techniques [1, 2].

Flash thermography (FT) provides a fast, safe, full-field, non-contact NDT approach, where a defect may be distinguished from the sound material through the mismatch of their thermal properties [3-5]. First, an intense optical flash is aimed at the inspected sample, which causes its surface temperature to rise rapidly. The induced thermal imbalance causes thermal waves to diffuse into the depth of the specimen. Meanwhile, the surface temperature is recorded using a high-precision infrared (IR) camera. At a hidden defect, the heat diffusion is hindered, which causes heat to accumulate above the defect. As such, defects can be identified as localized hotspots in the recorded thermographic sequence. However, due to the diffusive and highly damped nature of thermal waves, the reliable detection of small and deep defects is very challenging.

In order to obtain an improved defect detectability, the thermographic dataset is typically enhanced with image processing [6, 7] and/or data processing algorithms [5, 8-14]. Many of the data processing techniques operate either in temporal domain or in frequency domain (after applying a fast Fourier transform (FFT)). In this contribution, the defect detectability of several time- and frequency-domain processing techniques is evaluated. Several (industrial) carbon fiber reinforced polymer (CFRP) components, with a range of defect types, are inspected. The remainder of the paper is organized as follows: section 2 highlights the inspected samples and the experimental procedure. The considered post-processing techniques are briefly described in section 3, after which the results are discussed in section 4. Finally, section 5 concludes the paper.

2. Materials and Methods

The first CFRP component, depicted in Figure 1(a), is a 5.5 mm thick cross-ply $[(-45/0/45/90)_3]_s$ sample with five flat bottom holes (FBH) of diameter 20 mm at different depths, and is called $CFRP_{FBH}$ hereafter. Secondly, a sandwich part with woven CFRP skins and a Nomex honeycomb core is inspected (see Figure 1(b)). During production, a 30×30 mm^2 disbond was created between the skin and the core, while after curing, also a barely visible impact damage (BVID) was created on the same side. As a reference, the inspection result of vibrational testing is shown. The third inspected part is a component of the vertical stabilizer in an Airbus A320 (see Figure 1(c)), which has stiffeners at the back side. Three BVID damages were created by impacting the component with a 7.1 kg drop weight from three different drop heights (20 cm, 35 cm and 30 cm for $BVID_A$, $BVID_B$ and $BVID_C$, respectively). The vibrational NDT inspection results are shown as a reference.

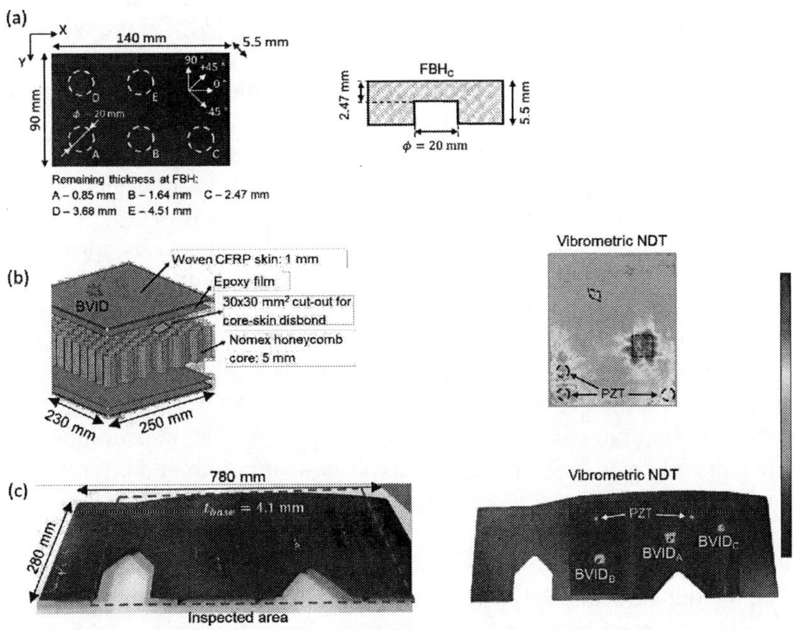

Figure 1: Inspected CFRP samples: (a) $CFRP_{FBH}$ with a schematic illustration of a flat bottom hole (FBH); (b) schematic illustration of $CFRP_{Sandw}$ and the result of a vibrometric NDT inspection; and (c) $CFRP_{Airbus}$ and the result of a vibrometric NDT inspection.

All samples are inspected with flash thermography in reflection mode, where a Hensel linear flash lamp (6 kJ, 5 ms flash) provided the optical excitation energy. A cooled FLIR A6750sc was used to record the surface temperature: 100 s at 20 Hz for the $CFRP_{FBH}$, 20 s at 30 Hz for the $CFRP_{Sandw}$, and 40 s at 30 Hz for the $CFRP_{Airbus}$. The IR camera has a focal plane array of 640×512 cryo-cooled InSb detectors, a noise-equivalent differential temperature difference (NEDT) of ≤ 20 mK and a bit depth of 14 bit. The camera is used in the calibration range of 10−90 °C, and is sensitive within the 3−5 μm mid- infrared range. The offset between the flash lamp and the inspected samples was ~300 mm for $CFRP_{FBH}$ and $CFRP_{Sandw}$, and ~500 mm for $CFRP_{Airbus}$. The distance between the IR camera and the $CFRP_{FBH}$ and $CFRP_{Sandw}$ samples was ~500 mm, and ~1000 mm for the $CFRP_{Airbus}$ component. Hard- and software modules from edevis GmbH guaranteed the accurate synchronization between the excitation and data acquisition. The obtained thermographic sequences are analysed in Matlab.

3. Post-processing Techniques

The data first undergoes a temporal standardization in order to suppress influences of non-uniform heating and other undesired effects [12, 14], after which it is further post-processed by several time- and frequency-domain processing algorithms.

In time domain, Dynamic Thermal Tomography (DTT) searches the value of the maximum thermal contrast and its associated moment in time, for every pixel. Therefore, we classify this here as a *single-bin evaluation* technique. The combination of these pixel values into images results in a $maxigram^T$ and a timegram, respectively [10, 14, 15]. On the other hand, Thermal Signal Area (TSA) integrates the thermal signal within a user-defined time interval [9], and is therefore categorized as an *integrated-bin evaluation* technique. It is preferential to integrate the temperature signal around the time at which a defect shows its maximum thermal contrast (which can be found with DTT) [9, 14]. However, since the same integration domain is considered for all pixels, this approach is not optimal if multiple defects at different depths are present (since they all have different moments of maximum thermal contrast).

The temporal data can also be transformed to frequency-domain data by applying a fast Fourier transform, after which the phase data is used since it has emissivity-normalized properties. The counterpart of DTT in frequency domain can be obtained by evaluating each pixel's maximum phase contrast and its corresponding frequency, and is called Frequency-Domain Tomography (FDT) [14]. These pixel values then combine to a $maxigram^F$ and a frequencygram. In frequency-phase domain, Adaptive Spectral Band Integration (ASBI) integrates the most-relevant spectral information (individually determined for each pixel without user input) in order to obtain a single damage index map [8]. ASBI is thus a frequency-domain *integrated-bin evaluation* technique.

4. Results and discussion

In this section, the defect detectability obtained with the post-processing techniques of section 3 is discussed for the three CFRP parts of section 2.

4.1. *Coupon with flat bottom holes ($CFRP_{FBH}$)*

The results of the different post-processing techniques on the $CFRP_{FBH}$ coupon sample are presented in Figure 2, with the true sizes and locations of the FBHs marked on the figures. Firstly, Figure 2(a,b) presents DTT's $maxigram^T$ and timegram. In the $maxigram^T$, only the three shallowest FBHs (0.85 mm, 1.64 mm and 2.47 mm depth) are detected. Additionally, there are clear effects of the fibre orientations which distort the background uniformity. The timegram in Figure 2(b) additionally hints the presence of the two deepest FBHs. The timegram is noisy since the values for sound pixels represent random fluctuations in the thermal contrast. Furthermore, the effect of lateral heat diffusion can be observed for the shallower defects, where the values in the timegram increase with the distance away from the FBH's centre. The defects therefore do not have a sharp outline in the timegram.

Several integration limits were carefully checked for TSA, and an integration over the first 50 s was found to provide a good detectability of the defects (see Figure 2(c)). This is a sensible integration range since it ranges over the timegram values of all FBHs (i.e. it encapsulates the times of maximum thermal contrast). In the TSA output, the three shallowest FBHs (0.85 mm, 1.64 mm and 2.47 mm depth) are clearly detected, while the presence of the fourth FBH (3.68 mm depth) is hinted. The background is significantly more uniform and less noisy than in DTT's maxigram^{T} and timegram. Of course, due to the integration procedure over a relatively long temporal range, the effects of lateral heat diffusion blur the edges of the shallow defects. In order to obtain better defect sizing, it would be beneficial to make the integration procedure adaptive in a pixel-wise manner. In that way, shallow defects will be associated with a shorter integration time which would limit lateral diffusion effects. Deep defects, on the other hand, will have longer integration times in order to increase their contrast.

The processed results of FDT are presented in Figure 2(d,e). The maxigram^{F} provides a good indication of the four shallowest FBHs (0.85 mm, 1.64 mm, 2.47 mm and 3.68 mm depth), and hints the presence of the deepest FBH (4.51 mm depth). Additionally, a highly uniform background is retrieved, which provides a significant improvement over DTT's maxigram^{T}. The frequencygram provides an even better indication of the deepest FBHs, but with a noisier background than the maxigram^{F}. Note that the frequencygram has the same values for the deepest defects due to the limited frequency resolution, which complicates the identification of their relative depth location. However, all defects have a very sharp outline in the frequencygram, as can be clearly observed in Figure 2 (e).

Lastly, ASBI's damage index map is shown in Figure 2 (f). It can be seen that all FBHs (depth up to 4.51 mm) are detectable with an almost uniform background and low noise level (note the logarithmic colour scale).

While DTT's maxigram^{T} and FDT's maxigram^{F} mainly serve for defect detection, DTT's timegram and FDT's frequencygram can be used for quantitative depth inversion. The main limitation for the timegram lies in its limited detectability of deep defects (FBHs detected clearly up to 2.47 mm depth, see Figure 2 (b)), however, quite uniform values are obtained over a defect's signature. On the other hand, the frequencygram can detect significantly deeper defects (detected up to 4.51 mm, see Figure 2 (e)), but the limited frequency resolution makes it practically very difficult to perform depth inversion for deep defects (deep defects have the same frequency values). With ASBI, only one single damage map is obtained, which serves for both defect detection and quantitative defect depth inversion (based on calibration curves from simulation) [8].

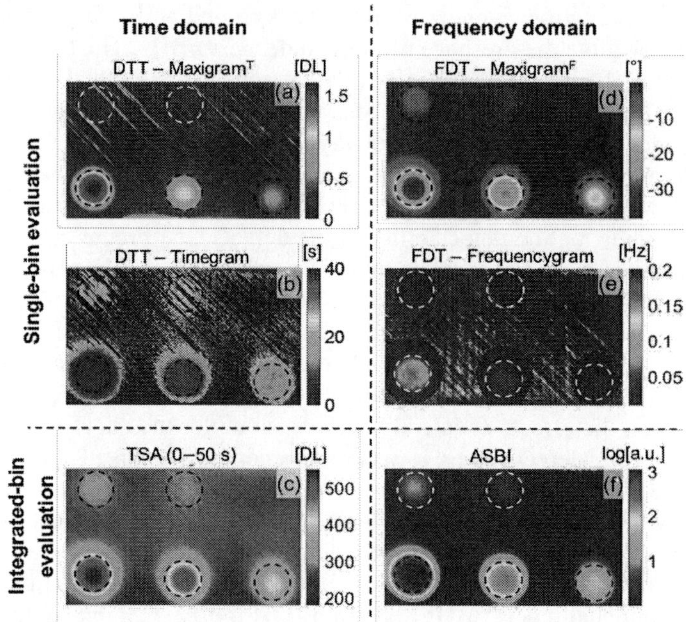

Figure 2: Processing results of the $CFRP_{FBH}$ sample: (a,b) DTT maxigram^T and timegram; (c) TSA (0-50 s); (d,e) FDT maxigram^F and frequencygram; and (f) ASBI damage index map.

4.2. *Sandwich sample with disbond and BVID ($CFRP_{Sandw}$)*

The post-processed results of the $CFRP_{Sandw}$ sandwich part are displayed in Figure 3. With DTT's maxigram^T (Figure 3(a)), only the shallow BVID damage is detectable, while no trace of the skin-core disbond is observed. On the other hand, the woven structure of the CFRP skins is clearly visible in the result. In the timegram (Figure 3(b)), a hint of the disbond is observed, however, the background noise makes its detection non-trivial. The noise also covers a wide temporal range which appears to hide the BVID, however, a rescaling clearly indicates the BVID's presence (see inset). For TSA, it was found that integrating over the first 10 seconds after the flash excitation provided a good defect detectability of both the BVID and the skin-core disbond (see Figure 3(c)). TSA does not provide a uniform background, which is somewhat disturbing for defect detection. The disturbed region in the bottom right corner originates from lateral heat diffusion effects of the sample's clamping.

The maxigram^F obtained by applying FDT (Figure 3(d)) provides a very clear detection of both the impact damage and the disbonded region. The BVID and the disbond are both also identified in the frequencygram in Figure 3(e). As expected, a comparison of the dominant frequencies for both defects reveals that the disbond is located deeper than the BVID (defect depth is inversely proportional to the frequency of maximum phase contrast). Lastly, both defects are also straightforward to pinpoint in ASBI's damage index map (see Figure 3(f)). Besides providing a higher defect detection quality than TSA, ASBI additionally obtains a more uniform background than TSA.

The horizontal disturbance in the results are remnants of the non-uniform heating profile (linear flash bulb), which were not completely suppressed by temporal standardization (see Figure 3(c)) nor by phase calculations (see Figure 3(d-f)).

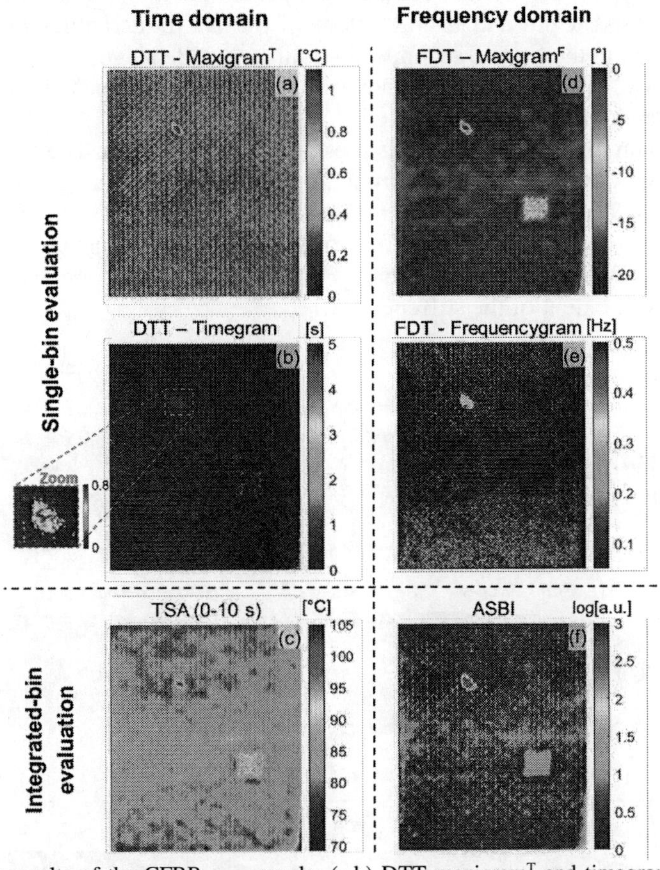

Figure 3: Processing results of the $CFRP_{Sandw}$ sample: (a,b) DTT maxigram^T and timegram (with zoom of the BVID area); (c) TSA (0-10 s); (d,e) FDT maxigram^F and frequencygram; and (f) ASBI damage index map.

4.3. *Aircraft component with 3 BVID damages ($CFRP_{Airbus}$)*

The post-processed results of the Airbus A320 component with three BVIDs are presented in Figure 4. Notice that in each of the subpanels, the clamp that was used to hold the sample is located between the two stiffeners on the right. At first sight, the background obtained with DTT's maxigram^T is very noisy (see Figure 4(a)). The vertical stiffeners at the back side are observed as locations of reduced maximum thermal contrast (due to the increased material thickness). The three impact damages can be detected, however, the strong background non-uniformity is a disturbing factor. DTT's timegram (Figure 4(b)) provides a comparable detectability as the maxigram^T, but with a less pronounced background non-uniformity. In this case, the stiffened regions are highlighted as areas of increased time values (i.e. later maximum thermal contrast) in comparison to the skin plate. It is interesting to observe that the two left BVIDs (which are not located on top of a stiffener), are surrounded by a thin border of higher time values (and lower maximum thermal contrast values, see the maxigram^T in Figure 4(a)), which indicates an increased depth and lateral extent of the BVID delaminations. The integration range corresponding to the first 10 seconds of cooling provided a good defect detectability with TSA (see Figure 4(c)). All three BVID's are easily detected as regions of increased value. Also the backside stiffeners are sharply indicated.

In frequency domain, FDT's maxigram^F is presented in Figure 4(d). The three BVID's, as well as the stiffeners, are easily found. Furthermore, a highly uniform background is obtained in the non-defected areas. The frequencygram in Figure 4(e) may also be used for the inspection of the features of this component, but suffers from a more pronounced background non-uniformity. Lastly, ASBI's damage index map (Figure 4(f)) provides a clear detectability of the defects, and in lesser extent of the stiffeners.

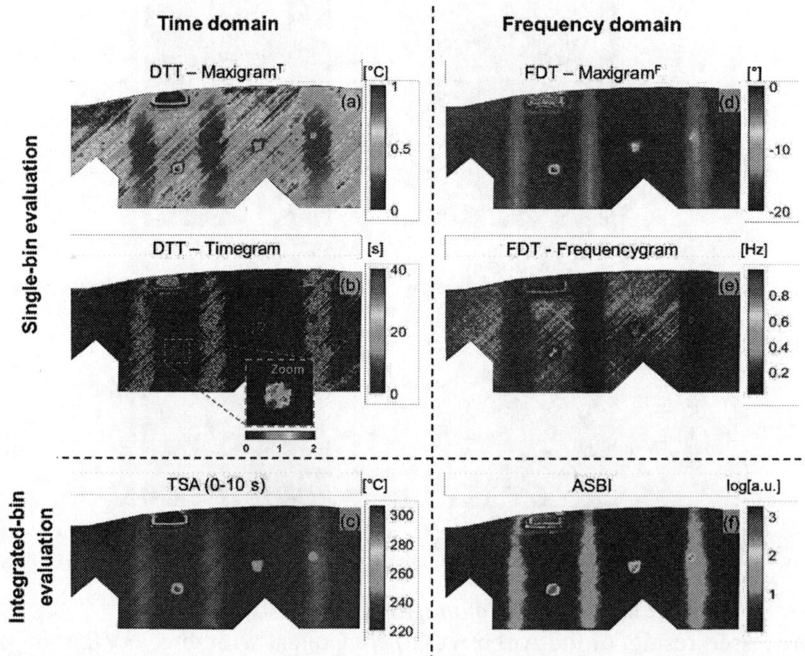

Figure 4: Processing results of the $CFRP_{Airbus}$ sample: (a,b) DTT maxigram^T and timegram; (c) TSA (0-10 s); (d,e) FDT maxigram^F and frequencygram; and (f) ASBI damage index map.

5. Conclusions

In this contribution, the performance of different post-processing techniques for flash thermography was evaluated in terms of their defect detectability. Both time- and frequency-domain, as

well as single-bin and integrated-bin, approaches are investigated. The study was performed on both monolithic and sandwich CFRP samples, and considered different defect types.
In general, it could be observed that the frequency-domain analysis techniques had a superior depth probing range than the time-domain techniques. They also provided an improved suppression of background noise. Especially the integrated-bin technique (ASBI) showed a high performance for the different studied cases.

Acknowledgements
The authors acknowledge the SBO project DETECT-IV (Grant no. 160455), which fits in the SIM research program MacroModelMat (M3) coordinated by Siemens (Siemens Digital Industries Software, Belgium) and funded by SIM (Strategic Initiative Materials in Flanders) and VLAIO (Flemish government agency Flanders Innovation & Entrepreneurship). The authors also acknowledge Fonds voor Wetenschappelijk Onderzoek-Vlaanderen (FWO-Vlaanderen) through grant 1S11520N. The authors further thank SABCA Limburg for providing the aircraft panel.

References

1. W. Nsengiyumva, et al., 'Advances, limitations and prospects of nondestructive testing and evaluation of thick composites and sandwich structures: A state-of-the-art review', Composite Structures: p. 112951, 2020.
2. M.R. Jolly, et al., 'Review of Non-destructive Testing (NDT) Techniques and their Applicability to Thick Walled Composites', Procedia CIRP, Vol 38: p. 129-136, 2015.
3. K. Chatterjee, et al., 'A comparison of the pulsed, lock-in and frequency modulated thermography nondestructive evaluation techniques', NDT & E International, Vol 44, No 7: p. 655-667, 2011.
4. J.M. Milne and W.N. Reynolds, 'The non-destructive evaluation of composites and other materials by thermal pulse video thermography', Thermosense VII: p. 119-1222, 1985.
5. V.P. Vavilov and D.D. Burleigh, 'Review of pulsed thermal NDT: Physical principles, theory and data processing', NDT & E International, Vol 73: p. 28-52, 2015.
6. C. Ibarra-Castanedo, et al., 'Infrared image processing and data analysis', Infrared Physics & Technology, Vol 46, No 1-2: p. 75-83, 2004.
7. G. Poelman, et al., 'Multi-scale gapped smoothing algorithm for robust baseline-free damage detection in optical infrared thermography', NDT & E International, Vol 112: p. 102247, 2020.
8. G. Poelman, et al., 'Adaptive spectral band integration in flash thermography: Enhanced defect detectability and quantification in composites', Composites Part B: Engineering, Vol 202: p. 108305, 2020.
9. M. Moradi and M.S. Safizadeh, 'Detection of edge debonding in composite patch using novel post processing method of thermography', NDT & E International, Vol 107: p. 102153, 2019.
10. V.P. Vavilov, 'Dynamic thermal tomography: Recent improvements and applications', NDT & E International, Vol 71: p. 23-32, 2015.

11. U. Netzelmann and D. Müller, 'Modified pulse-phase thermography algorithms for improved contrast-to-noise ratio from pulse-excited thermographic sequences', NDT & E International, Vol 116: p. 102325, 2020.
12. N. Rajic, 'Principal component thermography for flaw contrast enhancement and flaw depth characterisation in composite structures', Composite Structures, Vol 58, No 4: p. 521-528, 2002.
13. X. Maldague and S. Marinetti, 'Pulse phase infrared thermography', Journal of Applied Physics, Vol 79, No 5: p. 2694-2698, 1996.
14. G. Poelman, et al., 'An Experimental Study on the Defect Detectability of Time- and Frequency-Domain Analyses for Flash Thermography', Applied Sciences, Vol 10, No 22: p. 8051, 2020.
15. V.P. Vavilov and M.V. Kuimova, 'Dynamic Thermal Tomography of Composites: A Comparison of Reference and Reference-Free Approaches', Journal of Nondestructive Evaluation, Vol 38, No 1: p. 2, 2018.

NDT 4.0 – A Technology Overview

Lennart SCHULENBURG [1]

[1] VisiConsult X-ray Systems & Solutions GmbH, Brandenbrooker Weg 2-4,
D-23617 Stockelsdorf, Germany; Phone: +49 451 290286, Fax +49 451 29028622; l.schulenburg@visiconsult.de

Abstract
The world around us is changing at a rapid pace. Emerging megatrends like robotics, artificial intelligence, cloud computing and many more are reshaping activities in our personal and professional life. This presentation will address the major drivers of the digital transformation and how it affects our industry. It will identify disruptions, threats and opportunities that professionals, managers and companies have to address in order to be successful in the future. We will also spend some time looking on the impact of the COVID-19 pandemic on NDE and how the New Normal is affecting the way we have been working in the past. This presentation will help NDE experts to understand "the why" of the emerging NDE 4.0 transformation. Due to the expertise of the lecturer, the talk will heavily focus on Radiography (RT). To provide a better understanding there will be many practical examples that will illustrate the concepts.
Keywords: NDT 4.0, Automation, Robotics, Digitalization, Radiography

1. Introduction

NDE 4.0 is a commonly established term for a plethora of technologies. As with every innovation it is crucial to understand why companies and managers should implement them for their operation [1]. Therefore, one must understand the underlying economic drivers that lead to the necessity to change the status quo. The fact that NDE 4.0, like Industry 4.0, is a collective term for technologies ranging from Artificial Intelligence (AI), over robotics to Additive Manufacturing (AM) leads to an unclear picture of the relevance for individual stakeholders. Certainly, not all technologies provide value in the same way for every company and situation. Therefore, it is important to explore the technologies combined into NDE 4.0 individually. It is crucial that responsible managers carefully analyze the impact of each aspect on their operation and business. Is there a risk for an external disruption? Is it possible to increase efficiency? Is a fundamental process transformation required? As NDE managers are operating in a commercial setting, it is required to formulate the impact in economic terms. Of course, technical evaluation is also very important, and every responsible person must first check the technical viability and whether an innovation can be certified against industry standards. But finally, the innovation must be assessed in terms of economic viability. This means there must be an assessment of the costs, savings and ultimately Return of Investment (ROI). Otherwise, it is impossible to quantitatively justify the investment to the company's management or controllers.

2. NDE 4.0 in practice

The concept of NDE 4.0 is simply a subset of change sectors that are driven by the far bigger movement of Industry 4.0. It is important to understand where the term comes from. Figure 1 shows the different phases of industrial revolution and the next section will briefly describe the history of the different NDE industrialization stages on the example of Radiographic Testing (RT) [2].

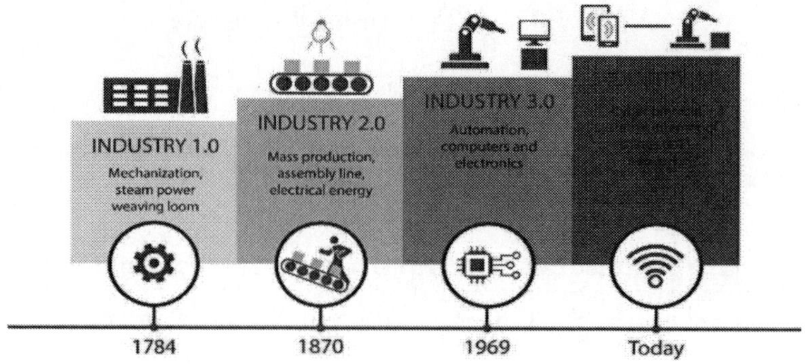

Figure 1: Development from Industry 1.0 to Industry 4.0 is comprised of gradual improvements in technology.

2.1 The Evolution of Industry 4.0

The first industrial revolution happened between the late 1700s and early 1800s. By using water and steam-powered engines in addition to other types of machine tools manufacturing evolved beyond the manual labor performed by people. In Radiographic Testing this phase is representing the status of using analog film and isotopes.

In the early part of the 20th century, the world entered the second industrial revolution with the use of electricity in factories increasing efficiency and productivty. Mass production concepts such as the assembly line were introduced. In Radiographic Testing the changes of this phase would be to incorporate electric components to improve the process and workflow. Particularly, this goes along with the usage of X-ray tubes that replace the isotopes, simple manipulators and automated film developers.

Starting in the late 1950s, a third industrial revolution slowly began to emerge, as manufacturers began incorporating more electronic and eventually computer based technology into their factories. A shift from analog and mechanical technology to digital technology and automation software started. Within the RT world this phase is characterized by the transition to digital detectors, NC programmable manipulators, digital archiving systems and robotics. This phase is often confused with the fourth phase. It is very important to note that a robot and an ERP connection is not enough to qualify for an NDE 4.0 system.

The fourth industrial revolution is characterized by connected computer systems, cloud utilization, the Internet of Things (IoT), Smart Factories, Big Data, Artificial Intelligence (AI), autonomous robots, predictive maintenance, and Additive Manufacturing (AM). The adoption of these technologies is leading to substantial improvements in productivity. Within the NDE world this results in fully integrated in-line systems, cloud connectivity, usage of AI for interpretation and advanced analytics. High performance gains were realized by the integration of NDE directly into the manufacturing line. Self-adapting systems allow automation to be used even on very small batch sizes. Defects are automatically detected and compared against inspection criteria and Computed Tomography (CT) is used to three dimensionally reconstruct objects and to perform complex analysis.

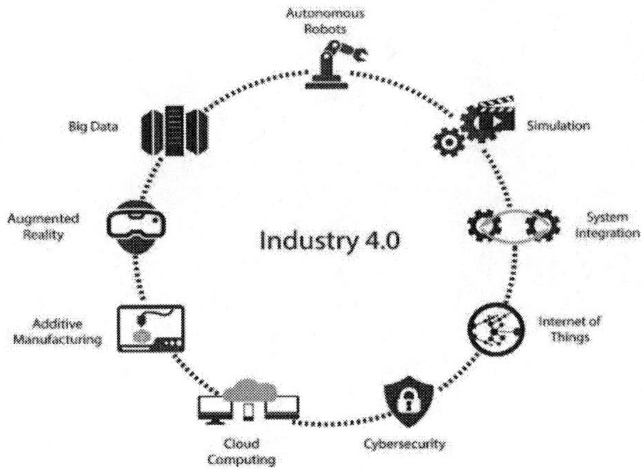

Figure 2: The different fields of Industry 4.0 and thereby NDE 4.0 are a collection of technology drivers.

This means that NDE professionals on the shop floor face fundamental challenges and may have a steep learning curve ahead as the new tools often require a different approach and perhaps even a completely different skill set [3]. Managing this change is the key for future success and competitiveness. Figure 2 shows the different fields of Industry 4.0, which also apply to our sector. This section will assess some of them and should provide a good overview, by introducing real examples where they already impact NDE today.

2.2 Robotics

First, we will investigate robotics and simulations, allowing repetitive handling tasks to be automated, allowing higher throughput, lower inspection costs and higher process safety. Figure 3 shows an example solution where three robots work in jointly to inspect airducts and pipes used in the aerospace industry. This team of robots collaborate, sharing the tasks of part handling and inspection. This way, cycle time is effectively reduced from several hours to a several minutes and a healthier work environment is created for operators, where heavy and potentially dangerous tasks are performed by machines [4].

Image quality is always supervised as the system performs automatic long-term performance evaluations according to ASTM E2737. New programs can be programmed offline – including the option to use a CAD/CAM simulation tool so that the system can be utilized 100% for production and does not need to be shut down for engineering purposes, thereby significantly increasing system utilization and throughput. To further optimize the process X-ray technicians can simulate the X-ray images digitally before even loading the part into the system. This allows to easily check the inspectability of the part and enables operators to establish the right X-ray parameters very early in the process. Systems these days have to communicate with the factory or enterprise mangement software (ERP/MES) and many other systems in the shop floor. The advantage of automated systems grows expontially if generated data can be used in real-time to fine-tune the manufacturing process.

Figure 3: Autonomous robots that are connected to a Manufacturing Enterprise System (MES) are working together to perform a digital X-ray inspection of complex welded pipes.

2.3 Artificial Intelligence, Big Data and Automated Quality Control

Another significant focus of Industry 4.0 includes Artificial Intelligence (AI) and Big Data. These concepts are realized for example as Automated Defect Recognition (ADR) in Radiography [5]. The R&D driven company VisiConsult X-ray Systems & Solutions GmbH for example, is currently developing an AI platform to detect defects even more reliably and more automated than in the past. X-ray technology is always used in non-destructive testing. Unlike optical processes, internal defects like pores, inclusions or geometries can also be tested. By replacing analogue film with digital detectors the data is available in real time and can be be evaluated and processed immediately.

For 20 years, the company has been delivering ADR systems, which detect errors automatically after a human did a component-specific parameterization. Specialists from VisiConsult predict that by using artificial intelligence, this effort could be reduced massively. Initial tests with available AI frameworks already achieved detection rates of up to 90% in a blind study compared to the conventional procedure.

Crucial for the success of an AI is the number and quality of the underlying image data. For this reason, VisiConsult has developed a cloud solution into which cooperation partners can upload evaluated image data. These are anonymized and used to parameterize the AI. The higher the quality of the data, the better the AI can become. VisiConsult's defect database includes multivalued multi-user records with marked and classified errors. As a specialist for fully automated inspection systems, VisiConsult has developed a patent for automatic robot positioning: Through the combination of robotics, 3D sensors and image registration a high positioning repeatability can be achieved without additional components. This system will now be combined with the new AI solutions, reducing the ramp-up process from several days to a few hours.

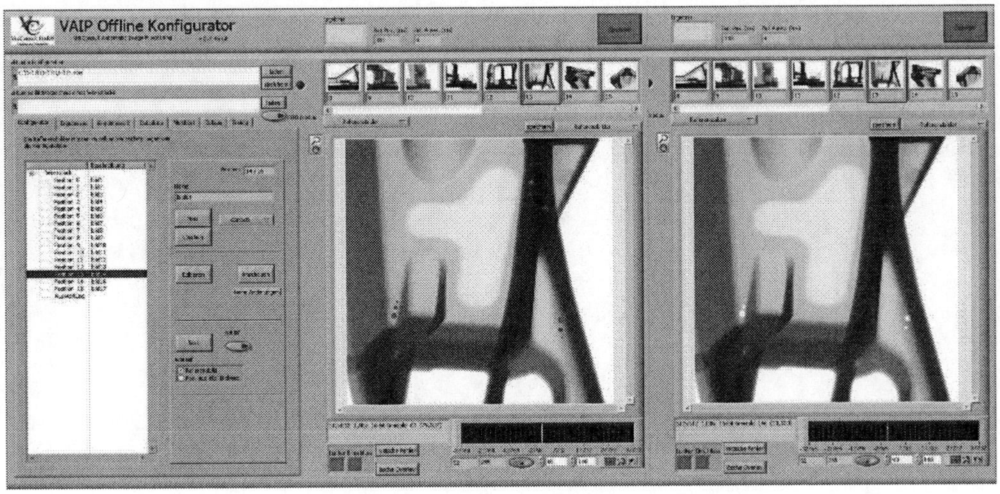

Figure 4: Result of an automated defect recognition (ADR).

2.3.1 Defect Recognition

Typical applications for defect recognition are casting parts. The ADR software should automatically classify typical casting defects like porosities and inclusions. Figure 4 shows the result of such a classification. On the left side is the evaluated part and defect list, while on the right one can see the original X-ray image. Different parts of the image allow different error thresholds. This can be defined by the Level III through inspection Region of Interests. Depending on the inspection guideline criteria for classification different parameter such as defect depth (contrast), distance to surface, defect size or defects per area can be calculated.

2.3.2 Geometric Evaluation

Even though most applications in NDT are looking for typical defects in produced parts it might be important to check for geometry, alignment, completeness or other aspects inside complex parts. Dedicated ADR software allows to perform these checks automatically. Figure 5 shows automatic evaluation of a flash-bang grenade. Checks performed are presence of the safety pin, length of the fuse delay, homogeneity of the explosive material, alignment of internal structures and much more. X-ray is the only technique allowing to perform these checks in a non-destructive manner.

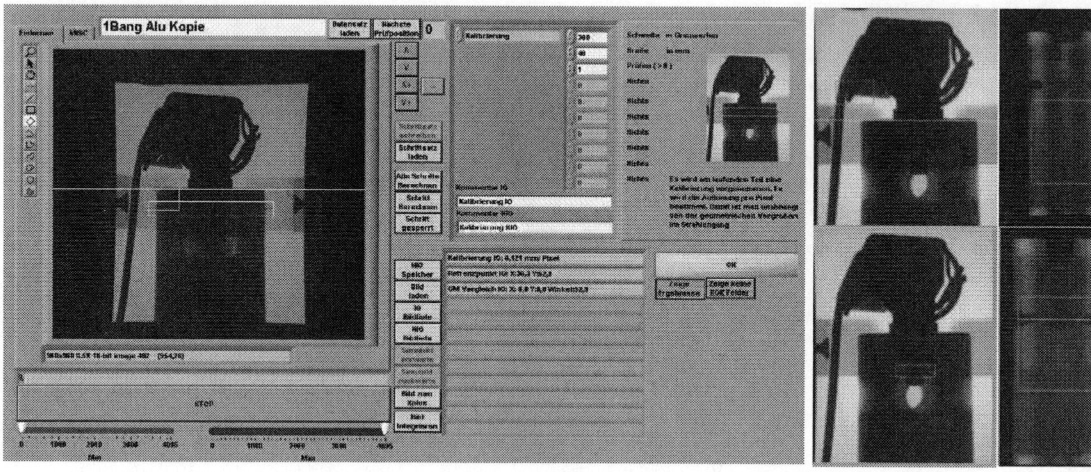

Figure 5. Automatic geometric measurements on a grenade.

2.4 Additive Manufacturing and Computed Tomography

Additive manufacturing (AM) offers new possibilities in manufacturing and designing products. The aerospace and automotive industry is a main driver because of the possibility to manufacturing lighter structures that reduce weight and save fuel. During the manufacturing process different discontinuities or defects can occur, depending on the applied AM technology. To ensure constant manufacturing quality of the parts it is required to perform a regular sampling or 100% inspection using non-destructing methods. Currently CT is the leading technology that is suited to sufficiently inspect complex AM parts to qualify them for safety critical environments [6]. By acquiring substantial amounts of digital radiographs and computing them into a 3D model, as can be seen in figure 6(a), we can gain information about parts like we never have before [7]. It is also possible to do advanced analyses like actual-nominal comparison, porosity analysis and metrology. With increasing computational power, we currently see CT moving from a lab environment to the shop floor. This allows to implement inline CT systems that perform a 100% inspection of parts while at the same time checking geometric tolerances.

The AM technology is growing unrestricted but there are some limitations include the standardization of the process, qualification of the parts and optimization of the manufacturing process. The motivation to use the opportunities of AM promote the development of qualified inspection techniques. CT offers a wide range of analysis methods to evaluate such complex structures depending on the application. In many places different institutes and organisations working on a standardisation for AM products and their qualification.

(a) 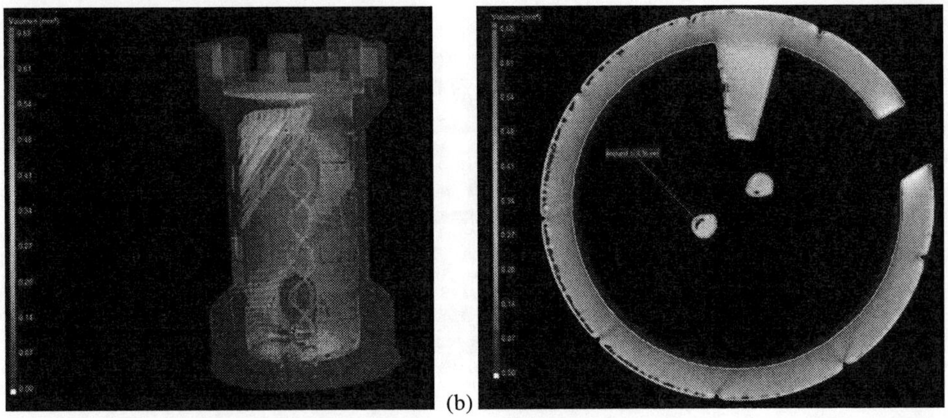 (b)

Figure 6: A CT-scan and porosity analysis of an AM part.

2.5 Digitalization – Cloud and Connectivity

With today's technology it becomes possible to connect nearly everything – system to remote viewer, systems to other connected systems, systems in different factories, and if desired even complete connected factories around the globe. Today it is possible to have image acquisition locally, computer scientists call it "the edge", and interpretation centralized in an excellence center. This could be a solution to compensate for the lack of qualified personnel that many companies face or temporary imbalances of workloads. Standardized interfaces on machine communication like OPC UA allow systems to interact, while standardized data formats like DICONDE [8] allow interoperability between devices from different manufacturers. Currently, NDE is often only seen to ensure quality, but it can provide so much more information.

VisiConsult offers a solution that can link systems and even global sites into a single inspection platform. It allows evaluations to be performed from remote locations or even from home office. This drastically reduces the need for highly skilled personnel at each individuall systems and at the same time it ensures a baseline for inspection quality. Figure 7 shows an exemplary setup using different modalities.

All data is archived in a local data center, and an AI is constantly trained in the background. The smarter the system gets, the more help the interpreters/inspectors get, which further improves the image evaluation quality, while the goal is to statistically prove that the probability of detection of the AI algorithms is good enough to be implemented. This simple scenario shows the powerful impact that robotics, cloud connection, system integration and AI can have on our industry.

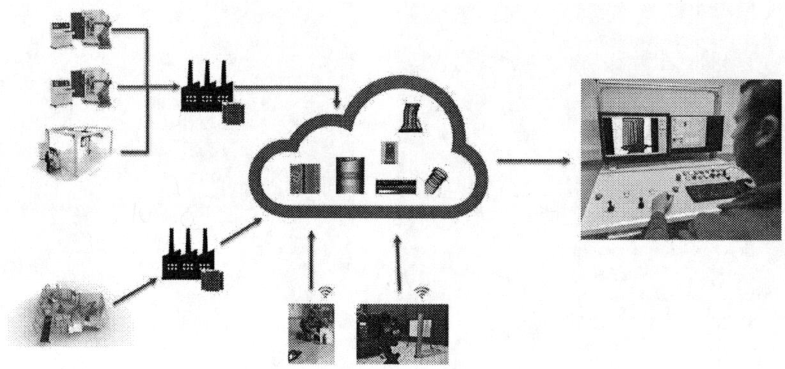

Figure 7: An example of a Cloud-network using a common evaluation platform.

3. The Impact of NDE 4.0

It is important to note that the value potential will differ between industries and individual companies [9]. In this section we will explore the different ways that NDE 4.0 can have a positive economic impact through improvements of efficiency and effectiveness. Additionally, we will also evaluate the possibility of disruptive innovations that create a completely new paradigm to value creation derived from NDE [10].

3.1 Incremental Innovation: Efficiency

Increasing the efficiency of processes has been the common theme of the past industrial revolutions. Moving from manual labor to steam powered machines (1.0) to manufacturing lines (2.0) to robotics, each industrial revolution has increased the efficiency of manufacturing and inspection processes. NDE 4.0 is also offering a huge potential for further efficiency increases.

The economic value of automation, and thereby efficiency increases, depends on the required investment costs and the operating cost reductions. To achieve the optimum cost the different factors must be carefully balanced. Ultimately, investments into process efficiency are most of the time depending on simple cost and volume considerations. The question is mainly, does the volume justify the investment and how will it influence the unit costs. It can very well be that a higher efficiency will indeed reduce the operating costs significantly, but the investment is spread out over so few units that it is not possible to recoup them over a reasonable period. There are many examples for efficiency increases: Implementation of process automation and robotics, digitalization of the data acquisition process and many more.

3.2 Incremental Innovation: Effectiveness

Innovations focused on increasing efficiency are quite straight forward. They take an established process and optimize it using process improvements or new technology. In some cases, the rigid quality process requirements in NDE make efficiency increases the only option to innovate. But often enough there is the possibility to also work on the process itself – to make it more effective. Typically, this type of innovation us is much harder to achieve as it requires to rethink the process and potentially also to perm a new process qualification. A reduced effective throughput might be caused for example by fluctuations in part throughput, unscheduled breaks, and other factors. In case of a X-ray imaging an aggregation of all images into a central data storage in the cloud and connect review stations from several locations might be considered to improve the effective throughput as shown in figure 7. Also, it can be assumed that more aggregated data leads to better trained ADR systems, which will further drive down the false reject rates and thereby saving costs. Increasing effectiveness follows the same economic principle as efficiency. Responsible managers must carefully analyze their operations to understand the composition of their cost curves. Where do costs come from (e.g. labor vs. consumables vs. CAPEX)? How easy can certain things be changed? What are the lowest hanging fruits and which levers have the biggest impact? When looking at the hard facts, it is very important to never forget the additional value that is added through soft factors. But even more important is to assess the risk that is associated with the innovation and whether the organization has the capabilities to implement it.

3.3 Disruptive Innovation: Paradigm Shift

The data-driven trends of Industry 4.0 and NDE 4.0 are fueled by information provided from industrial sensors – also called the Industrial Internet of Things (IIoT) [11]. These sensors are omnipresent and range from discrete data points such as temperature, time, position, etc. to the most complex sensors that provide a much higher depth of information. One such sensor found in NDE are for example CT systems – arguably a sensor that has the potential to deliver a high level of information regarding a company's product and processes. In manufacturing in general, one of the hottest technologies with high potential is the concept of digital twins, which are considered as virtual copies of manufactured products or equipment. What people sometimes do not realize is that NDE can supply a lot of information that can be used to build this digital twin! Imagine the possibilities when you have all the as-built data for every component combined with ever-advancing simulation capabilities. This is a paradigm shift for our industry, transforming NDE from a pure quality step to a full industrial sensor in the manufacturing process. So we must ask ourselves what else can we learn from the hundreds and thousands of images that we are generating? The possibilities are vast, and the NDE world can be proud to provide that information. The step to see NDE not only as a Quality Assurance (QA) and Quality Control (QC) process, but as a comprehensive process sensor, is a revolutionary paradigm shift [9].

4. Conclusions

This paper has shown some of the key-aspects of Industry 4.0 within NDT. It shows that we are already on the journey towards implementation in our industry. Are we there yet? For sure not. The main challenges are that we do not have a complete conversation from analogue (film) to digital operation principles. This third industrial revolution is still ongoing – in some industries faster in others slower.

Another challenge is training and education. The new technology requires a completely different skillset than the technology of the past. It is highly suggested to start early to train the needed skills to responsible operators. We can typically see a level of rejection and fear if automation is in discussion. Even the best AI or automated systems will not replace trained and skilled operators.

When assessing the potential of innovations, it is recommended to take a step back and to assess the effectivity ("Are we doing the right things?") and efficiency ("Are we doing things right?"). Then one should proceed to analyze on which dimension the innovation will have an impact. This will help a lot to determine the right way forward and what it will take to get implemented. To perform an orderly risk assessment, it is helpful to classify whether it is an incremental or a disruptive innovation. Performing these analyses early will help to identify all stakeholders and parties that must get involved at an early stage [12]. Only when all stakeholders bought into the idea and the economic feasibility has been confirmed, one should take it to management for consideration.

To put it in a nutshell, NDT 4.0 is the movement of our century. Even though we have many challenges to overcome the benefits and gains of this technology will be massive. It will also be an enabler technology that will increase the added value of NDT to the production.

References

[1] R. Singh, "Singh, R.: Purpose and pursuit of NDE 4.0.," *Materials Evaluation,* vol. 78, p. 784–793, 2020.

[2] L. Schulenburg, "Schulenburg, L.: NDT 4.0—opportunity or threat?," VisiConsult X-ray Systems & Solutions GmbH, 2019. [Online]. Available: https://www.linkedin.com/pulse/ndt-40-opportunity-threat-lennart-schulenburg/.

[3] "Intelligence augmentation and human–machine interface best practices for NDT 4.0 reliability," *Materials Evaluation,* vol. 78, p. 869–879, 2020.

[4] H. Saboonchi, D. Blanchette and K. Hayes , "Advancements in Radiographic Evaluation Through the Migration into NDE 4.0," *J Nondestruct Eval 40, 17,* 2021.

[5] VisiConsult, "Automated Defect Recognition (ADR) – Definition and Application," 2019. [Online]. Available: https://visiconsult.de/technology/automated-defect-recognition/.

[6] L. Schulenburg and F. Herold, "Inspection of Additive Manufactured Parts using Computed Tomography," *The NDT Technician,* vol. 4, no. 1, pp. 1-5, 2021.

[7] N. Meyendorf, L. Bond, S. Curtis-Beard, S. Heilmann, P. Severi, R. Schallert, H. Scholz and C. Wunderlich, "NDE 4.0—NDE for the 21st Century— The Internet of Things and Cyber Physical Systems Will Revolutionize NDE," Singapore, 2017.

[8] ASTM E2339-15, Standard Practice for Digital Imaging and Communication in Nondestructive Evaluation (DICONDE), 03.03. ed., West Conshohocken: Book of Standards Volume, 2015.

[9] J. Vrana, "NDE Perception and emerging reality: NDE 4.0 value extraction," *Materials Evluation,* vol. 78, p. 835–851, 2020.

[10] R. Link and N. Riess, "NDT 4.0—overall significance and implications to NDT," Gothenburg, 2018.

[11] B. Valeske, A. Osman, F. Römer and R. Tschuncky, "Next Generation NDE Sensor Systems as IIoT Elements of Industry 4.0," *Research in Nondestructive Evaluation,* vol. 31, no. 5-6, pp. 340-369, 2020.

[12] J. Vrana and R. Singh, "A Design Thinking Perspective," *J Nondestruct Eval 40,* 2021.

Analysis of Ballistic Impacts on Composite materials by Infrared Active Thermography.

Stephane BOUBANGA TOMBET, Eric GUYOT

Telops 100-2600 St-Jean-Baptiste ave,
G2E 6J5, Québec, Québec, Canada ; Phone: +1 418-864-7808; e-mail: stephane.boubanga@telops.com

Abstract
We investigated the assessment of the damaged area on composites ballistic plates subjected to high velocity impact. Active pulsed thermography technique was used for performing post-mortem analysis of the impacted specimens. We conducted quantitative defect analysis combined with inputs of the velocity of the projectile, the analysis of the absorbed energy and differentiation of defects type. to evaluate how efficient the material is at spreading the absorbed energy to a large area. Our quantitative analysis of the damage area of the central impact show consistent results with the size of the projectile testifying of the good precision of the quantification done in this work. High velocity shoots appear to induce predominantly fiber breakage while and low velocity shoots induce predominantly delamination and matrix cracking.

Keywords: Ballistic Impacts, NDT, Active Thermography, defect size, High speed Infrared Camera

1. Introduction

Since the beginning of the last century the ongoing advances in materials engineering have led to an unrestrained development of new technologies. Composite materials are the ones attracting most attention because they have many advantages over their homogeneous counterparts. These include high specific stiffness and high specific strength combined with a significant reduction in weight [1-2] making them attractive for many industrial applications. One of the most important fields of application is the defense industry were the composites properties such as low weight, rigidity, strength and durability are of key importance. Composite materials made from artificially obtained high strength fibers [3] are particularly interesting. These composites are characterized by many fiber-reinforced properties that make them ideal for ballistic protection applications. The ballistic protection equipment should protect the user from for instance arms fire. The proper analysis strategy of the area of internal damage caused by the impact of bullets is very important in the research and evaluation of protective composite ballistic equipment. Damage to the internal structure of the composite coating material can only be assessed using non-destructive testing methods. Such methods, such as infrared thermography test methods, are particularly effective in the case of composite materials. This work focused on the assessment of the damaged area on composites ballistic plates subjected to high velocity impact. Active pulsed thermography technique was used for performing post-mortem analysis of the impacted specimens. The post-mortem analysis was combined with inputs of the velocity of the projectile, the absorbed energy to evaluate how efficient the material is at spreading the absorbed energy to a large area.

2. Experimental information.

2.1 Ballistic impact testing

In this study bore gas gun was used to propel the projectile into the specimen. The selected projectile is a 17 grain fragment simulating projectile (FSP). The 16 in by 16in ballistic plate was rigidly held by a square metallic holder and position only a few inches in front of the gun muzzle. Projectile velocity and yaw are recorded from a perpendicular high-speed camera operating at 25 000 frame per second. The ballistic impacts on the specimen are classified in 3 categories: Perforation, generated at high projectile speed, partial perforation (in red), and bouncing generated at low projectile speed (see Figure 1).

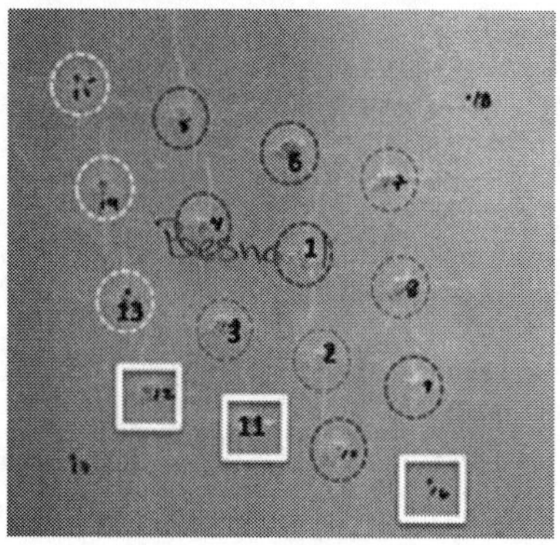

Shot # [-]	Type [Perforation/Partial/Bouncing]	Projectile Speed [m/s]
8	Pe	566
7	Pe	548
3	Pe	545
6	P	523
2	Pe	521
4	P	480
1	P	454
5	P	449
9	P	421
10	P	409
13	B	335
14	B	300
15	B	262

Figure 1. Composite Ballistic sample used for NDT measurements (left) and shots description and projectile speed (right).

2.2 Active thermography testing.

Active thermography experiments were conducted with Telops new non-destructive testing solutions called TESTD. Flash lamp source with pulse energy of 6 KJ was used to excite the wooden laminate sample after the ballistic testing. Telops high-definition Infrared camera was used to capture the sample cooling after the pulse heating and Fourier transform analysis were conducted to obtained phase images (see Figure 2).

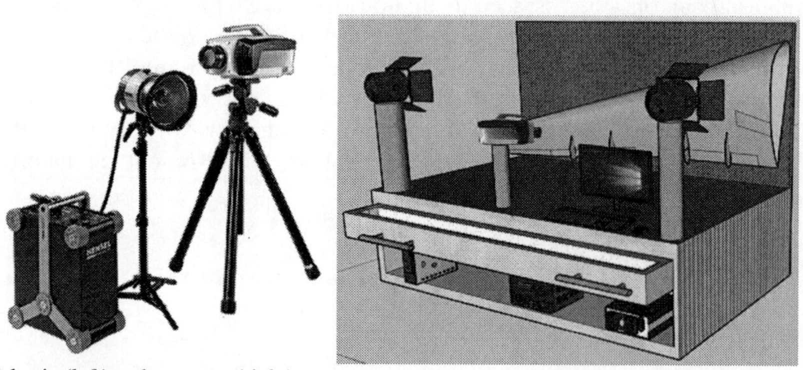

Figure 2. Telops TESTD basic (left) and compact (right) systems.

3. Results.

We analyzed three different impact type: bouncing, perforation and partial perforation at different projectile speeds. Figure 3.a depicts a phase image of the impacted ballistic composite sample. The defects are identified by the contrast in the image resulting from their relative location in the material volume. During ballistic impact the main energy absorption mechanisms include projectile kinetic energy imparted onto the sample, and energy absorption as a result of shear plugging, tensile fiber failure, fiber debonding, fiber pull-out, matrix cracking, interlaminar delamination, and frictional energy absorbed during interaction of the bullet and laminate. The damage modes observed for the investigated composite sample are detailed in Fig. 3.a.

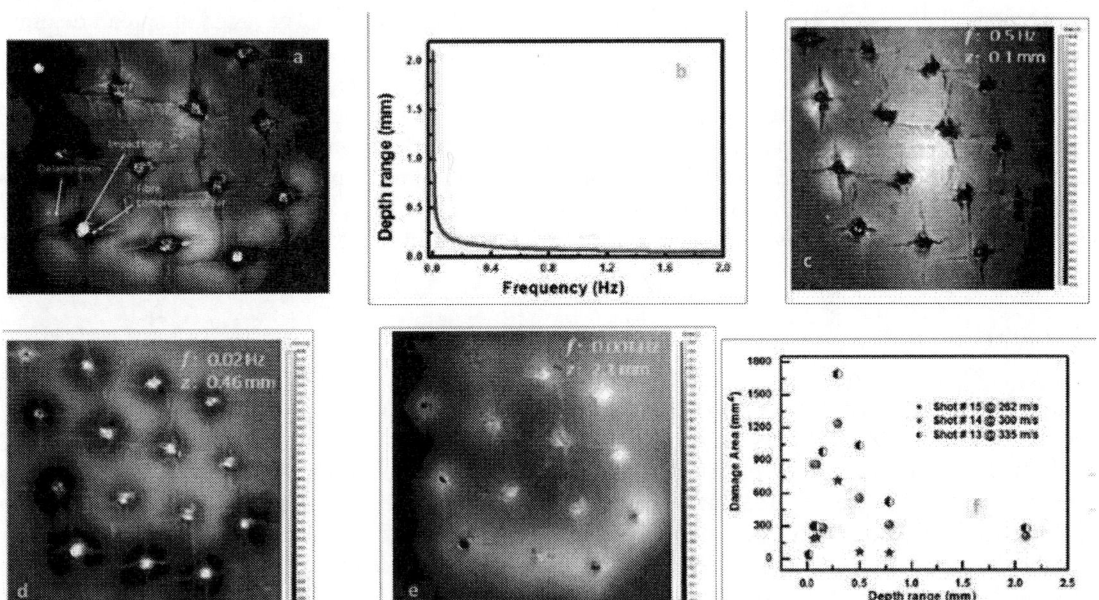

Figure 3. Phase image with different type of defects (a). Depth Range for the investigated composite sample (b) and 3 phase images at different depths within the sample (c-e). Damage area for three shots (bouncing) as a function of depth range (f).

Impact damage at the surface of the sample can be seen and quantified using the high frequency phase image, while the low frequency phase image is used to assess the subsurface/internal damage. The relationship between the frequency and the investigated depth in the sample volume is shown in Fig. 3b, while figures 3c, 3d, 3e show phase images of the ballistic plate at different frequencies and the impact damage modes at different depth in the plate. We Quantified each damage size and figures 3f the total damage area for three shots (bouncing type) as a function of the depth range. The total damage includes the impact hole and regions displaying the fiber delamination and compression/shear induced damage type. Higher velocity shots appear to induce greater damage. Bouncing type shots induce delamination very close to the sample surface but sample damage size quickly decreases with increasing depth. For example, shot #15 does not show any damage at 2.1 mm from the surface of the sample. This allows for an estimation of the depth of the impact damage for Shot #15 15 to be approximately 1 mm. In the case of partial perforation shots (velocities from 409 m/s to 523 m/s) we observed relatively higher delamination damage was observed for lower speed shots (shots 9 and 10). However, for some shots, the projectile remained embedded in the sample after the ballistic impact test, this could have had an undetermined impact on the NDT results. Additional considerations include the effect of sample weakening due to repeated impacts. The sample was subjected to 15 consecutive shots, later shots impacted a sample already weakened by earlier shots. Considerations must therefore be made for impacts recorded in the latter phases of this test to account for potential degradation of overall plate integrity due to previous impacts. Additionally, in this analysis, shots 9-12 generated increased and overlapping areas of delamination damage compared to earlier impacts, potentially due to the plate degradation by earlier shots. These large areas of spread delamination damage complicate efforts to precisely quantify the damage area induced by each shot. In the case of perforation shots (velocities from 521 m/s to 566 m/s) we observed that the defect sizes do not vary much with the projectile speed as expected in high velocity ballistic impact conditions. Compared to the partial perforation and bouncing shots, the perforation shots show less delamination damage. When the projectile completely perforates the material more energy is absorbed via the fiber breakage mechanism. Relatively limited delamination damage is therefore observed with corresponding smaller defect size as compared to the partial perforation and the bouncing type.

4. Conclusions

When sample is impacted by the projectile, at the moment of impact the projectile is slowed by a large number of individual fibers. As a result of the impact, fibers in the sample stretch and break to absorb the kinetic energy of the projectile casing. This creates a subsurface defect in the composite structure with a much greater area than the caliber of the projectile. We showed in this work that it is possible to assess the degree of destruction of the internal material. We conducted quantitative analysis of the damaged areas on composites ballistic plates subjected to high velocity impact. This quantitative defect analysis combined with knowledge of projectile velocity allows for characterization of absorbed energy and differentiation of generated defect types. Our observations indicate that high velocity shots tend to induce smaller impact damage areas characterized primarily by fiber breakage, while low velocity shots tend to induce larger impact damage areas featuring predominantly delamination and matrix cracking damage mechanisms.

References

1. Hull D, Clyne TW (1996), An introduction to Composite Materials, Cambridge University Press, Cambridge.

2. Jones RM (1975), Mechanics of Composite Materials, Hemisphere Publishing Corporation, New York.
3. Reid SR, Zhou G. Impact behaviour of fibre-reinforced composite materials and structures. Woodhead Publishing; 2000.

Magnetic Barkhausen Noise Analysis for Microstructural Effects Separation during Flow Forming of Metastable Austenite 304L

Julian ROZO VASQUEZ [1], Bahman ARIAN [2], Markus RIEPOLD [3]
Werner HOMBERG [2], Ansgar TRÄCHTLER [3], Frank WALTHER [1]

[1] Chair of Materials Test Engineering (WPT), TU Dortmund University, Dortmund, NRW, Germany;
Phone: +49 231 755-8535, Fax: +49 231 755-8029; e-mail: julian.rozo@tu-dortmund.de; frank.walther@tu-dortmund.de
[2] Chair of Forming and Machining Technology (LUF), Paderborn University, Paderborn, NRW, Germany;
e-mail: ba@luf.uni-paderborn.de; wh@luf.uni-paderborn.de
[3] Fraunhofer Institute for Mechatronic System Design (IEM), Paderborn, NRW, Germany;
e-mail: markus.riepold@iem.fraunhofer.de; ansgar.traechtler@iem.fraunhofer.de

Abstract
This work deals with the analysis of the influence of different microstructural effects on non-destructive micromagnetic testing using magnetic Barkhausen noise (MBN) analysis, during flow forming of austenitic steel 304L tubes. This is necessary for the implementation of non-destructive MBN testing within the model of a soft sensor for closed-loop control systems. A characterization of the specimens was carried out by means of α'-martensite measurements, hardness tests, residual stresses and surface roughness. The aim is to establish the influence of the different microstructural effects on the measured MBN parameter, the maximum MBN amplitude M_{max}. The strain-induced hardening, evolution of residual stresses and phase transformation during plastic deformation have a direct influence in the development of the M_{max} parameter. However, a lower surface roughness triggers a mitigating effect on the MBN signals. This information is valuable to define the operation window of the sensor within the material modeling for property control.

Keywords: Magnetic Barkhausen noise, microstructural effect separation, ferromagnetic phase, α'-martensite, flow forming, metastable austenitic steel, soft sensor, closed-loop control

1. Introduction

The implementation of control engineering within metal forming processes has gained recently relevance in the production of high quality components with optimized properties in industries like automotive and aeronautics. Metastable austenitic stainless steels have been used in these applications thanks to their high strength, corrosion resistance and good formability. During plastic deformation of these steels, strain-induced phase transformation of non-magnetic austenite into ferromagnetic α'-martensite occurs [1]. This phase transformation can be measured by means of the change of magnetic properties using the non-destructive magnetic Barkhausen noise (MBN) effect [2,3]. Non-destructive testing plays a key role in the implementation of control engineering in production systems, since the measured parameters can be used as control variables of the closed-loop control systems.

In this work, non-destructive testing, specifically MBN analysis, has been used to monitor the strain hardening and phase transformation during flow forming of austenitic steel AISI 304L. This entails the formulation of correlations between the measured MBN parameters and different microstructural phenomena that occur during manufacturing processes. Many studies describe correlations between MBN testing and hardening [4], residual stresses [5], surface condition [6,7] and phase transformation [2,8], that occur during plastic deformation of steels. Since the MBN measurements contain the effects of the mentioned phenomena, the identification of their influence is crucial to model the material behaviour. The focus of this work is the assessment of the influence of those factors on the measured MBN signals on flow formed components. This is helpful to define the operation window of the micromagnetic sensor in the frame of the soft sensor design for closed-loop property control.

2. Materials and Methods

2.1. Specimens

The specimens were manufactured using seamless tubes made of stainless steel AISI 304L (X2CrNi18-9, 1.4307) with 80 mm outer diameter and 4 mm wall thickness. The production was carried out by means of the flow forming machine BD 40 (Bohner & Koehle GmbH & Co. KG, Esslingen, Germany) (Figure 1a) at the LUF. The specimens were manufactured at room temperature, with a rotation speed of s = 30 rpm and using feed rates between f = 6 and 60 mm/min. The focus will be on the specimens produced with feed rates of 6 mm/min (specimen 1) and 60 mm/min (specimen 2). The flow forming operation was performed using three rollers to reduce the wall thickness of the tubes. The initial 4 mm wall thickness was reduced by 1 mm in each flow forming step. Starting from the initial condition "IC" (metastable austenite), different forming operations were performed to produce three forming zones (FZ 1, 2 and 3) (Figure 1b). Each forming zone corresponds to a plastic deformation state with a defined amount of strain-induced α'-martensite. The final workpieces manufactured during the production process are shown in Figure 1c.

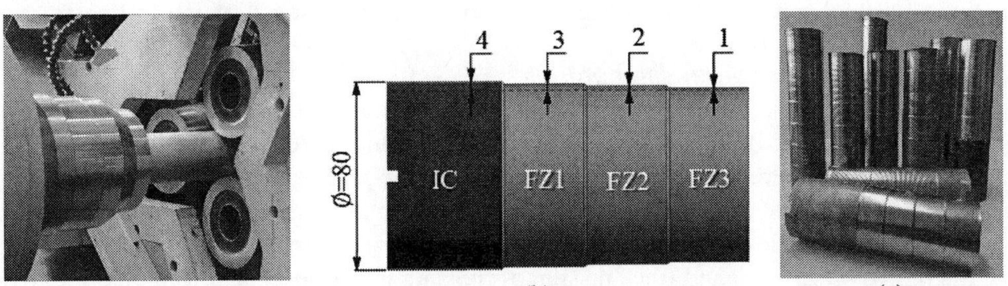

(a) (b) (c)

Figure 1. Specimen manufacture: (a) Flow forming machine; (b) geometry and dimensions (in mm), (c) final workpieces.

2.2. Measurements of surface roughness and microscopic inspection of the surface condition

The roughness of the initial condition and the different forming zones of the specimens 1 and 2 were measured by means of the tactile device MarSurf M300 (Mahr GmbH, Goettingen, Germany) (Figure 2a). The values of the arithmetic mean roughness (Ra) and the mean peak-to-valley height (Rz) of the outer surface of the specimens were measured. The data was obtained locating the probe tip normal to the curved outer surface of the specimens, using the corresponding prisms (Figure 2a).
In order to obtain graphical information of the surface condition, an inspection was performed using the digital microscope Keyence VHX-500F (Keyence Corp., Osaka, Japan) (Figure 2b).

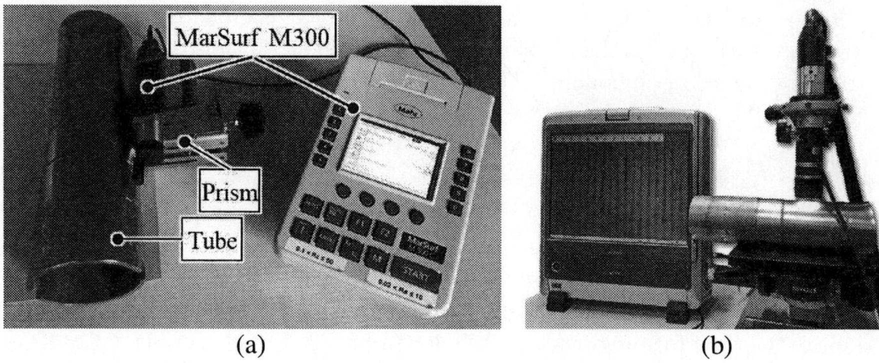

Figure 2. (a) Roughness measurements with the MarSurf M300 and (b) microscopic inspection of the surface with the Keyence VHX-500F.

2.3. Analysis of residual stress by means of X-ray diffraction (XRD)

The residual stresses were determined by means of the analysis of X-ray diffraction of the outer surface of the specimens. The measurements were carried out using the X-ray diffractor Bruker D8 Discover (Bruker Corp., Billerica, Massachusetts, USA) with a 1 mm collimator. The analysis of residual stresses for the austenitic phase (IC) was performed at $2\theta = 95.989\ ^\circ$, which corresponds to the Miller index {h k l} = {2 2 2}. Similarly, the analyses of the martensitic phase (FZ1 to 3) were carried out at $2\theta = 82.329\ ^\circ$ in the plane {h k l} = {2 1 1}. The data to perform X-ray diffraction analyses regarding the mentioned 2θ angles were extracted from the report by Eigenmann et al. [9].

2.4. Measurements of α'-martensite content

Measurements of the α'-martensite content close to the outer surface of the tubes were carried out by means of the Feritscope FMP30 (Helmut Fischer GmbH, Sindelfingen, Germany). This device works under the magneto-inductive principle to measure the ferritic phase content. By means of the correlations developed by Talonen et al., 2004, the α'-martensite contents can be determined from the ferritic phase measurements [8]. However, these measurements cannot be used in closed-loop control systems, because the online acquisition and transmission of the data is not possible during the production process. For this reason, the use of micromagnetic testing is a convenient alternative as soft sensor for the property control.

2.5. Micromagnetic testing by means of the 3MA-II system

Micromagnetic analyses are based on the behavior of ferromagnetic materials, such as metastable austenitic steel AISI 304L. In ferromagnetic materials, the magnetic induction B, triggered by a magnetic field H, can be described by means of the magnetization hysteresis (Figure 3a). These materials are divided by means of Bloch walls in spontaneously magnetized regions, called magnetic domains. During magnetization and demagnetization of the material, the walls move in response to the applied field. This motion is jerky and discontinuous, producing the Barkhausen effect, that can be evidenced in the magnetization hysteresis at high scale. This effect is influenced by the interaction of the walls with lattice defects, dislocations, phase changes, grain boundaries and residual stresses. This way, from the magnetic Barkhausen noise (MBN) curve (Figure 3a), the maximum amplitude of the MBN profile (M_{max}) is the parameter of interest [10–12].

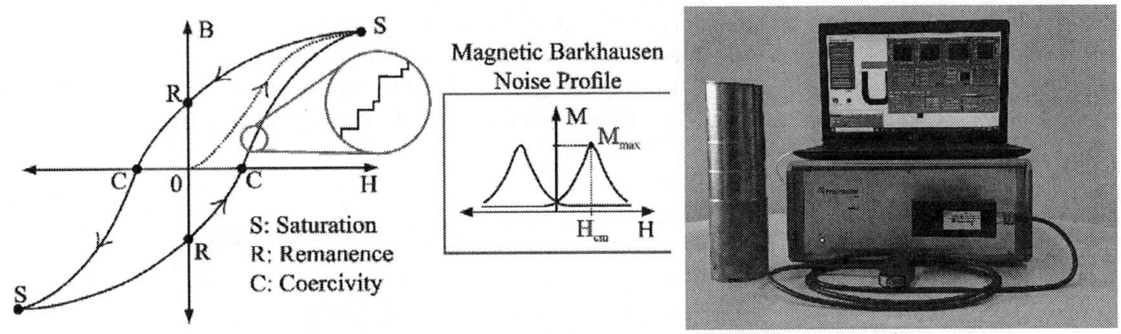

(a) (b)

Figure 3. Micromagnetic testing: (a) Hysteresis loop and MBN profile; (b) 3MA-II system device and flow-formed specimen.

In this work, the micromagnetic measurements were carried out using the 3MA-II system (Fraunhofer IZFP, Saarbruecken, Germany), with the sensor SN18201 in contact with the outer surface of the specimens (Figure 3b).

3. Results and Discussions

3.1. Surface condition inspection and roughness measurements

An inspection of the surface of both specimens were performed with the digital microscope Keyence VHX-500F. The results reported in Figure 4 correspond to the forming zone 2 (FZ2) of both specimens. The upper pictures correspond to a 30x magnification and the lower to 100x. No remarkable difference of the surface condition was evidenced on the different forming zones of the same specimen. A qualitative analysis of the results evidences a higher degree of surface damage in the specimen produced with a feed rate f = 60 mm/min (specimen 2). For a constant rotation speed s = 30 rpm of the workpiece, the higher the feed rate of the roller tools, the lower the contact time with the outer surface of the specimen. This entails a less uniform plastic deformation evidenced in lower surface quality.

A quantitative validation of this inspection was carried out by means of roughness measurements. The results for specimens 1 and 2 and their respective forming zones are summarized in Table 1. The measurements were performed as described in section 2.2 using the tactile device MarSurf M300. The roughness values of the initial condition in both specimens are similar, with values around Ra = 1.33 µm. The mean roughness values of the specimens produced with 6 mm/min (specimen 1) are 90% lower than those for specimens produced with 60 mm/min (specimen 2). Since the rotation speed remains constant, the contact time between the roller tools and the workpieces is higher at lower feed rates, as a result the plastic deformation is more uniform.

The roughness values of the different forming zones (FZ1 to FZ3) within the same specimen show a decreasing behaviour. A higher number of overruns is required to achieve FZ3 compared to FZ1. This entails a higher contact time between the roller tools and the workpiece, therefore a more uniform plastic deformation is evidenced in zones with higher number of overruns. An analysis of the data for specimen 2 shows that the best roughness value (in FZ3) is higher than the roughness of the initial condition, which is undesired concerning the surface quality of the components. This allows the definition of process windows regarding the parameters for production of optimized components.

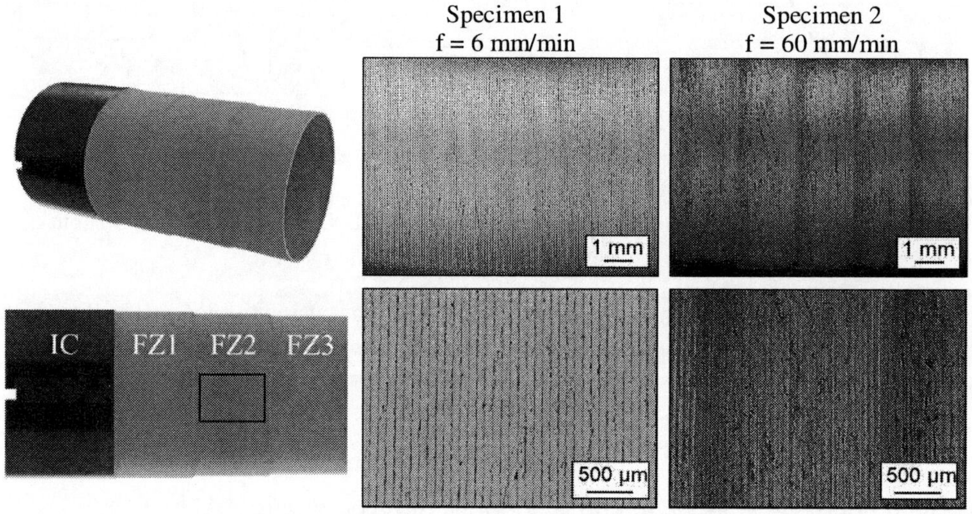

Figure 4. Microscopic investigation of the surface condition of specimens produced using different feed rates.

Table 1. Results of surface roughness measurements for the different forming zones of specimens produced with two different feed rates

Specimen	Forming zone	Surface roughness	
		Arithmetic mean roughness, Ra [µm]	Mean peak-to-valley height, Rz [µm]
Specimen 1 f = 6 mm/min	IC	1.33	10.03
	FZ1	0.321	2.599
	FZ2	0.343	3.318
	FZ3	0.249	1.718
Specimen 2 f = 60 mm/min	IC	1.38	10.5
	FZ1	3.47	13.63
	FZ2	2.449	11.76
	FZ3	1.961	8.555

The values of surface roughness of both specimens on the different forming zones were plotted in Figure 5 to identify their influence in the non-destructive testing by means of the micromagnetic analysis.

3.2. Residual stresses

The residual stresses of the outer surface of the flow formed specimens were measured according to the specifications described in section 2.3. The results (Table 2) show that the workpieces have compressive residual stresses in the initial condition and the different forming zones. At the initial condition the measured residual stresses were around -547.5 MPa. The plastic deformation in both specimens causes an increment of the residual stresses, reaching a maximum about -220 MPa. The successive plastic deformation on FZ 2 and 3, increases the compressive residual stresses to obtain values -374.3 MPa for specimen 1 and -326.4 MPa for specimen 2. It is well known that compressive residual stresses induced by means of manufacturing processes like turning and in this case metal forming, increase fatigue life, corrosion and wear resistance [13]. This entails a positive effect of the flow forming of tubes for their final uses.

Table 2. Results of residual stresses for the different forming zones of specimens produced with two different feed rates

Specimen	Forming zone	Residual stress, RS [MPa]
Specimen 1 f = 6 mm/min	IC	-547.5
	FZ1	-229.8
	FZ2	-365.1
	FZ3	-374.3
Specimen 2 f = 60 mm/min	IC	-547.5
	FZ1	-220.1
	FZ2	-285
	FZ3	-326.4

The values of residual stresses were plotted in Figure 5 to analyse their effect in the non-destructive micromagnetic testing.

3.3. Separation of effects during non-destructive testing by means of micromagnetic analysis

Figure 5 shows the correlation of the non-destructive micromagnetic testing by means of the magnetic Barkhausen noise (MBN) analysis with other characterization techniques. The graphics show the evolution of different microstructural phenomena that occur during flow forming of austenitic stainless steel 304L. The strain-induced phase transformation during plastic deformation is shown by means of the α'-martensite content. The microstructural mechanisms of phase transformation during flow forming were discussed by the authors in [14]. The higher the thickness reduction, the higher the α'-martensite volume. The strain-induced hardening is well correlated with the phase transformation. Likewise, the maximum amplitude of the MBN profile (M_{max}) grows consequently with the evolution of α'-martensite volume and the hardness, as it was evidenced in previous investigations [2,15]. This parameter is even sensible to the saturation at the final states of deformation (FZ2 and 3) for the specimen 1, as can be evidenced in the plateau of the curves.

A comparison of the α'-martensite volume and the hardness in both specimens leads to the conclusion that specimens produced with lower feed rates, undergo a faster saturation and higher values of α'-

martensite. This is related to the discussion on section 3.1 regarding the uniformity of plastic deformation depending on the feed rate parameter. As discussed in section 2.5, the MBN analysis contains a mixture of different microstructural effects that occur during plastic deformation like phase transformation, grain boundaries, residual stresses and surface damage, evaluated here by means of surface roughness. The transition from the non-magnetic austenite to the ferromagnetic martensitic phase is the predominant effect influencing the growth of M_{max} with the deformation. This phase transformation entails a microstructural change in the grain size, which increases the amount of grain boundaries and consequently the number of pinning sites of the magnetic domains [2,10]. For this reason, higher MBN values correspond to higher deformation states, according to the references [16,17]. As it was described in section 3.2, the compressive residual stresses in the martensitic phase (FZ1 to 3) increase with the deformation. This entails an increment of M_{max}, which is consistent with previous investigations [5,18].

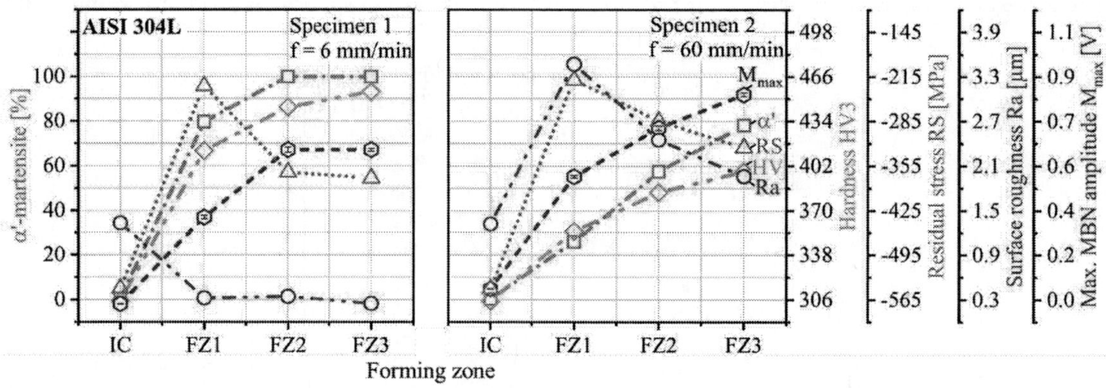

Figure 5. Results of the correlations between non-destructive testing by means of MBN and microstructural phenomena like hardening, residual stresses and roughness, during plastic deformation of austenitic steel AISI 304L.

However, one controversial aspect arises from the fact that the M_{max} curve of the specimen 1 develops below to the corresponding curve of the specimen 2. Despite higher α'-martensite content in specimen 1 in all deformed states (FZ1 to FZ3), the measured M_{max} parameter is always higher for the specimen 2. An analysis of the data in Figure 5 allows to conclude that the good surface condition of specimen 1 after plastic deformation has an attenuating effect on the micromagnetic measurements performed with 3MA-II System. This result is consistent with previous studies that prove the suitability of the non-destructive micromagnetic testing to characterize wear and surface damage. In those investigations the more wear the surfaces have, the higher the measured M_{max} values [6,19].

This information is valuable to define the operation window of the micromagnetic sensor for property control. The soft sensor model to define the α'-martensite content in terms of the variable M_{max}, should consider the remarkable influence of the surface condition of the workpieces on the micromagnetic measurements.

3. Conclusions and outlook

In this work, an assessment of the influence of different microstructural phenomena on non-destructive micromagnetic testing by means of magnetic Barkhausen noise (MBN) analysis was successfully carried out. The information obtained is valuable to define the operation window of the micromagnetic sensor in a closed-loop property control during flow forming of austenitic stainless steel tubes. Plastic deformation of this material triggers residual stresses, strain-induced hardening and phase transformation, specifically from metastable austenite into α'-martensite. It was proved that the MBN testing can be used to monitor the microstructural evolution of the specimens during deformation, because there is a direct correlation between the measured MBN signals and the content of the ferromagnetic phase α'-martensite. Consequently, the increment of compressive residual stresses matches with the incremental behavior of the micromagnetic parameters. However, it was evidenced that the quality of the tested surfaces plays a key role in the evolution of the MBN signals. A better surface quality causes an attenuation of the micromagnetic measurements. As the surface quality depends on process parameters like feed rate, a comprehensive assessment of the connection between process parameters and component properties was carried out. This means a step forward in the soft sensor model for a closed-loop property control.

Future work will focus on the implementation of these results within the material model of the soft sensor for property control. This entails a quantitative assessment of the surface condition effect on micromagnetic testing. With this information, an improvement of the soft sensor is possible, since the operation process window of the micromagnetic sensor will be related to the process parameters and the final goal to produce high quality products with optimized properties.

Acknowledgements
The authors would like to thank the German Research Foundation (Deutsche Forschungsgemeinschaft, DFG) for the support of the depicted research within the priority program SPP 2183 "Property Controlled Deformation Processes", through project No. 424335026 "Property Control During Spinning of Metastable Austenites".

References

[1] P. Haušild; K. Kolařík; M. Karlík, 'Characterization of strain-induced martensitic transformation in A301 stainless steel by Barkhausen noise measurement', Materials & Design, Vol 44, pp 548–554, 2013.

[2] M. Astudillo; M. Nicolás; J. Ruzzante; M. Gómez; G. Ferrari; L. Padovese; M. Pumarega, 'Correlation between martensitic phase transformation and magnetic Barkhausen noise of AISI 304 steel', Procedia Mater. Sci., Vol 9, pp 435–443, 2015.

[3] J. Anglada-Rivera; L. Padovese; J. Capó-Sánchez, 'Magnetic Barkhausen Noise and hysteresis loop in commercial carbon steel: Influence of applied tensile stress and grain size', J. Magn. Mater., Vol 231, pp 299–306, 2001.

[4] M. Neyra Astudillo; N. Núñez; M. López Pumarega; J. Ruzzante; L. Padovese, 'Magnetic Barkhausen noise and magneto acoustic emission in stainless steel plates', Procedia Mater. Sci., Vol 8, pp 674–682, 2015.

[5] S. Schuster; L. Dertinger; D. Dapprich; J. Gibmeier, 'Application of magnetic Barkhausen noise for residual stress analysis – Consideration of the microstructure', Mater. Test., Vol 60, No 6, pp 545–552, 2018.

[6] L. Samfaß; N. Baak; R. Meya; O. Hering; A. Tekkaya; F. Walther, 'Micro-magnetic damage characterization of bent and cold forged parts', Prod. Eng., Vol 14, No 1, pp 77–85, 2020.

[7] M. Teschke; J. Rozo Vasquez; L. Lücker; F. Walther, 'Characterization of damage evolution on hot flat rolled mild steel sheets by means of micromagnetic parameters and fatigue strength determination', Materials, Vol 13, No 11, 2020.

[8] J. Talonen; P. Aspegren; H. Hänninen, 'Comparison of different methods for measuring strain induced α-martensite content in austenitic steels', Mater. Sci. and Tech., Vol 20, No 12, pp 1506–1512, 2004.

[9] B. Eigenmann; E. Macherauch, 'Röntgenographische Untersuchung von Spannungszuständen in Werkstoffen. Teil III', Matwiss. Werkstofftech., Vol 27, No 9, pp 426–437, 1996.

[10] D. Jiles, 'Introduction to magnetism and magnetic materials. Chapman & Hall, London, 1991, ISBN 978-0-412-38640-4.

[11] B. Cullity; C. Graham, 'Introduction to magnetic materials. IEEE/Wiley, Hoboken N.J., 2009.

[12] S. Chikazumi, 'Physics of magnetism. John Wiley & Sons, Inc., New York, NY, USA,, 1966.

[13] Edoardo Capello, 'Residual stresses in turning: Part I: Influence of process parameters', J. Mat. Proc. Tech., Vol 160, No 2, pp 221–228, 2005.

[14] J. Rozo Vasquez; B. Arian; M. Riepold; W. Homberg; A. Trächtler; F. Walther, 'Microstructural investigation on phase transformation during flow forming of the metastable austenite AISI 304. In: Neidel, A. (Hrsg.) - Fortschritte in der Metallographie- Sonderbände der Praktischen Metallographie zur 54. Metallographie-Tagung, 75–81, Saarbruecken, Germany, 2020.

[15] M. Neslušan; J. Šugárová; P. Haušild; P. Minárik; J. Čapek; M. Jambor; P. Šugár, 'Barkhausen noise emission in AISI 321 austenitic steel originating from the strain-induced martensite transformation', Metals, Vol 11, No 3, pp 429, 2021.

[16] S. Yamaura; Y. Furuya; T. Watanabe, 'The effect of grain boundary microstructure on Barkhausen noise in ferromagnetic materials', Acta Mater., Vol 49, pp 3019–3027, 2001.

[17] Z. Wang; Y. Gu; Y. Wang, 'A review of three magnetic NDT technologies', J. Magn. Mater., Vol 324, No 4, pp 382–388, 2012.

[18] F. Eichhorn; M. Zschau, 'Messung des Eigenspannungsabbaus durch Analyse des Barkhausen-Rauschens', Z. Werkstofftech., Vol 11, pp 213–216, 1980.

[19] J. Tenkamp; M. Haack; F. Walther; M. Weibring; P. Tenberge, 'Application of micro-magnetic testing systems for non-destructive analysis of wear progress in case-hardened 16MnCr5', Mater. Test., Vol 58, No 9, pp 709–716, 2016.

A Simulation Platform for Structural Health Monitoring

Fabrice FOUCHER[1], Bastien CLAUSSE[1], Roman FERNANDEZ[1], Olivier MESNIL[2], Arnaud RECOQUILLAY[2]

[1]EXTENDE, 14 Avenue Carnot, 91300 Massy, France,
e-mail: fabrice.foucher@extende.com

[2]Université Paris-Saclay, CEA, List, F-91120, Palaiseau, France

Abstract

The CIVA platform is a well-known multi technique simulation and analysis software in NDT. Developed by CEA LIST, but also resulting from the contribution of numerous industrial and academic partners within Europe, the software keeps evolving to offer its users the best performances in terms of numerical efficiency, imaging, and reliability demonstration. CIVA can simulate UT, GWT, ET, RT, CT, and now Thermography in NDT. CIVA has also recently expanded its modelling tools to Structural Health Monitoring based on guided ultrasonic waves.

Based on a combination of spectral Finite Elements methods and a parametric macro-mesh technique, CIVA SHM can address many industrial SHM configurations (metallic or composite materials, planar, cylindrical, or curved geometries) with very competitive performance.

SHM by guided waves needs to rely on more quantitative performance demonstration and optimization studies to significantly increase its industrial deployment. This is why an efficient and dedicated simulation tool looks particularly useful since computational and hardware costs are very often prohibitive for large simulation campaigns with most traditional Finite Elements software packages.

After a brief description of the numerical techniques involved in this tool, some validation and application examples are presented in this paper.

Keywords: Simulation, Modelling, SHM, Guided Waves, Imaging, Performance demonstration, Optimization, MAPOD.

1. Introduction

Among Structural Health Monitoring methods, guided wave SHM is a technique relying on the permanent integration of a transducers' network aiming at detecting potential damages affecting the structural integrity of the monitored structure. To be part of a competitive predictive maintenance strategy, the choice of relevant GW-SHM sensors, their number, and their location in the specimen is of paramount importance to optimize critical flaws detection while limiting their number and thus the whole cost. Since building monitored and flawed prototypes is even more costly than simple mock-ups, simulation can play a key role as a virtual prototyping tool before physical implementation.

To increase the industrial deployment of GW-SHM systems, a solid performance demonstration stage is also required. Indeed, due to the fully automated measurement process, the large number of influential parameters involved, the numerous sources of false alarms and the potential failures of the embedded sensors themselves, reliability studies must be conducted. Experimental campaigns become rapidly prohibitive if you want to take all uncertainties into account. This is why simulation can help since it can provide a large amount of data and information regarding some of the most influential parameters, which will greatly reduce the need for costly mock-ups, limiting their number to a relevant sampling.

By nature, GW-SHM provides complex signals (several dispersive modes, several sensor paths, influence of component features and edges, etc.) and simulation can also be useful to help the interpretation of these results and develop relevant signal or image processing strategies. Simulation should help to improve our mastering of the SHM technique as well as generating large data sets to train deep learning algorithms in the context of an on-going Machine Learning approach of SHM analysis.

2. Modelling techniques and tools for SHM

CIVA is a software platform dedicated to NDE and SHM, widely used worldwide for the simulation and analysis of several NDE methods (Ultrasounds, Guided Waves, Radiography, Eddy Current and now Thermography). SHM by guided waves has recently joined this platform. Finding the best compromise between computation performance, easiness of use and versatility for the targeted application has always been the three pillars of the CIVA strategy.

Contrary to some other applications based on long range ultrasonic guided waves such as pipe NDE inspections where 2D-like models are suitable for a lot of configurations, modelling SHM applications requires to fully consider the three-dimensional nature of the wave propagation. However, traditional 3D Finite Elements approaches often fail to simulate realistic configurations due to prohibitive computational costs in an industrial context, that is why the choice has been made to rely on the Spectral Finite Elements technique (SEM) in CIVA SHM. The SEM technique implemented in CIVA relies on high-order elements coupled with an optimized meshing strategy thanks to a structured macro-mesh parameterized with respect to the geometry of the SHM configuration. For readers interested in the numerical techniques, more detailed information is provided in [2]. It has shown very competitive performances with accurate predictions obtained with a gain of a factor higher than 100 in terms of computation time compared to generic FEM packages and a negligible RAM footprint [3] which allows to use CIVA SHM on a traditional hardware or launch several simulations in parallel on a cluster.

An industrial tool for SHM also requires convenient user interfaces and relevant analysis environments. In terms of user interface, CIVA SHM benefits from the same NDE-oriented environment than the other modules of the CIVA platform. The mesh being fully parametrized, it is automatically managed without need for advanced FEM skills from the user. Due to the numerous sensors involved and the automatic monitoring process, SHM intrinsically generates a large amount of data, and it is challenging to be able to treat them efficiently. From the data collection, a data analysis stage is then required to provide meaningful metrics so that the decision-making process can follow. Imaging is one way to improve defects detection and identification by providing a clear signature for a given indication and CIVA SHM includes such imaging reconstruction techniques. Parametric studies, metamodels and scripting approaches are also compatible with CIVA SHM to help conducting efficient, massive, and customized sensitivity studies and analyses.

CIVA is developed by CEA and distributed worldwide by EXTENDE [1]. CIVA SHM can currently simulates metallic or composite panels, potentially multi layered, and with an Omega stiffener. An irregular curvature can be applied to the initial planar profiles so that components such as leading edges or bended parts can be modelled. Metallic cylinders can also be addressed. While CIVA lets you easily define typical piezo electric circular patch sensors assuming a radial uniform loading, user can also simulate other profiles such as a normal loading or a custom one. Sensors can be regularly or arbitrary positioned on the monitored structure. Provided results include the raw signals obtained on all receivers as well as wave field snapshots or local stress measurements but also, as mentioned above, defect signature reconstruction images through a Delay And Sum (DAS) algorithm [6], analogous to the Total Focusing Method in ultrasound imaging.

Many evolutions are in process for implementation in future releases, for instance new specimen geometries such as multi-layered and composite tubes or elbows, but also the accounting of attenuation during wave propagation, new sensor types such as Fiber Bragg Grating as well as additional imaging algorithms and the capacity to load experimental data.

3. Validation

As for any simulation software, it is important to rely on experimental validation references to verify the relevance and accuracy of the models and use it with confidence in an industrial project. Several validation studies have been conducted around CIVA SHM for isotropic metallic plates, cylinders, or for composite panels [3], [4]. Some of these validation cases are compared to experimental data provided by the « Open Guided Waves » initiative, detailed in [5].

Let's focus on one of these latter cases which deals with the instrumentation of a composite panel made with 16 carbon-epoxy plies for a 2mm total thickness. This type of structure exhibits anisotropic properties with respect to the ultrasonic guided waves propagation.

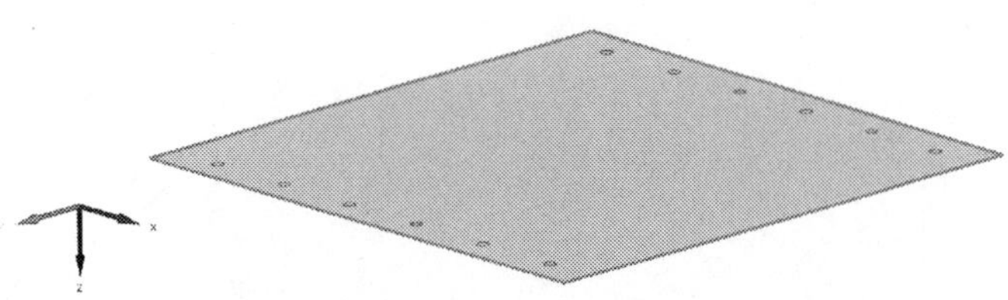

Figure 1: Monitored composite panel proposed in the Open Guided Wave validation benchmark.

The dispersion curves calculated by CIVA SHM are presented below in Figure 2. A0 mode data appears in red (group velocity of 1280m/s and wavelength of 19mm at 40kHz in direction 0°) while S0 is in blue (group velocity of 3400m/s and wavelength of 85mm at 40kHz in direction 0°). This composite material being anisotropic, a polar representation of group velocities versus the angle of the wave path is also provided.

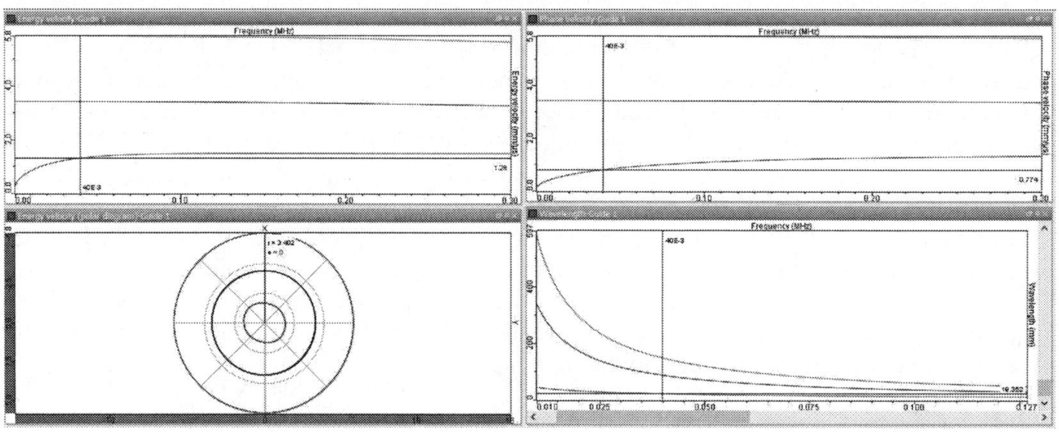

Figure 2: Dispersion curves calculated for A0 (red), S0 (blue) and S1 (green) modes (on the left: group velocity with a polar view at the bottom and a frequency domain view for the selected angle at the top; on the right: wavelengths at the bottom and phase velocity at the top).

12 piezo-electric patches of 10mm diameter are located as shown in Figure 1. Several excitation frequencies are considered, all sensors are successively excited as transmitters (with a radial load) while the 11 others are receivers (also called round-robin). In Figure 3 below, measurements and simulations are compared for the whole set of receivers when sensors #1, then #3, and then #8 (illustrated with a red point on the specimen schematic) are used in transmission with a 40kHz 5 cycles burst excitation. On the left, a B-Scan is shown where each column represents the signal received by one sensor (Simulated B-Scan is below the corresponding experimental B-Scan). On the right, each A-Scan signal is shown (simulation in red, experiment in blue). Experimental and simulated data are normalized with respect to the measured/simulated amplitude maximum on all the receivers channels (excluding the Pulse-Echo channel).

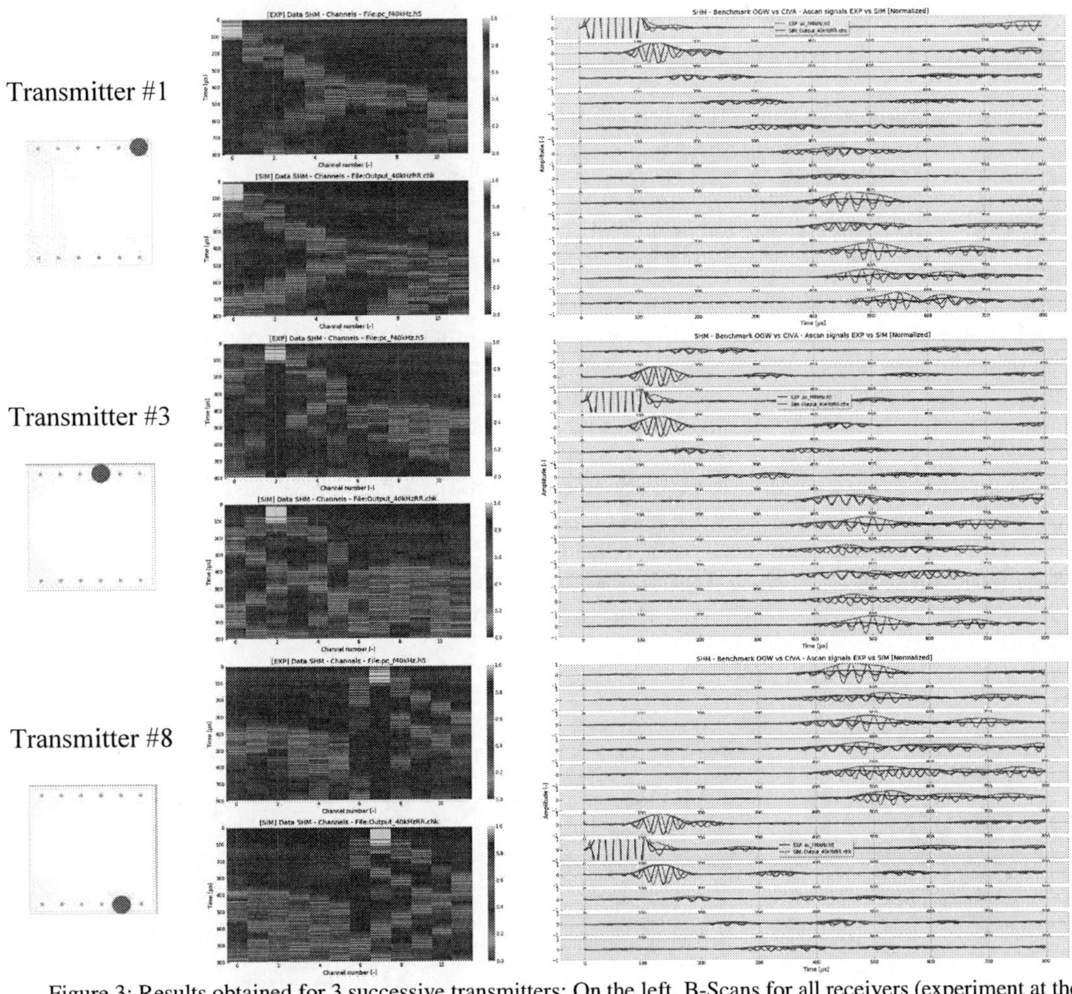

Figure 3: Results obtained for 3 successive transmitters: On the left, B-Scans for all receivers (experiment at the top, simulation at the bottom); On the right, detailed A-Scans (simulation in red, experiment in blue).

The comparison shows a good agreement between the calculated signals and the measurements: the different modes are well predicted (first the fastest S0 mode, then A0 one), the time of flights, the signals shapes as well as the different signals components due to the reflections on the edges appear very representative of the measurements. Modelling is never an exact copy of a measurement, due to both the model approximations and the experimental uncertainties. For the case under study, the main uncertainties are linked to the reflections on the specimen edges (assumed to be perfect in the simulation model), especially because the sensors are close to the plate edges. It can be also mentioned that the wave attenuation is currently neglected in the simulation model. Despite this, the agreement between simulation and experiment looks very satisfying to validate the use of CIVA SHM on this application case to predict the transmitted modes and the interactions with the specimen and potential defects. More details on these validation cases are given in [4].

4. Imaging applications for defect detection and identification

Defects detection is based on the processing of the signals received by the sensors' network which usually relies on the comparison between the defective state and a reference « baseline » state. While this process can be directly performed on the time domain raw signals with subtraction and signal processing techniques, imaging tomographic reconstruction techniques also exist to provide a more comprehensive information to detect and localize an anomaly. Of course, guided wave imaging will be efficient only if the generated modes are suited to the defect to inspect (in terms of wavelength mainly). Moreover, the imaging technique can also produce artefacts due to the algorithm itself or due to parasitic signals (for instance those due to the reflection on specimen edges or features).

The composite panel and sensors network (radial excitation) previously described are studied below in Figure 4. Each image displayed below results from DAS processing of signals predicted without any flaw and with a through-hole of 20mm diameter located in at the specimen center. Here three configurations are considered with similar sensors/flaw relative positions but with three different plate dimensions: 500mm*500mm (validation case), 700mm*700mm and 1000mm*1000mm.

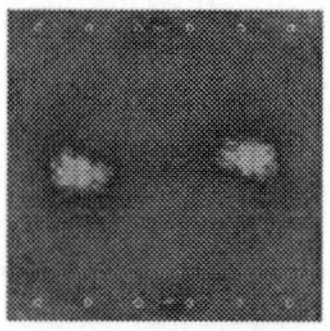

Figure 4a: Hole Ø20mm Plate 500mm*500mm.

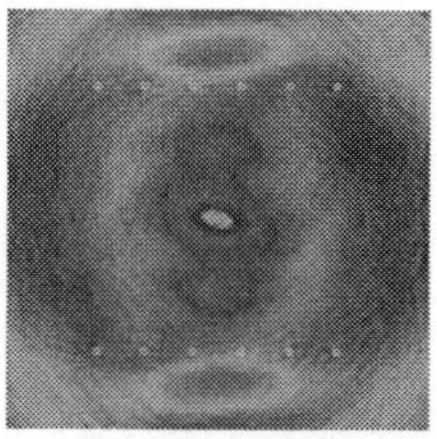

Figure 4b: Hole Ø20mm Plate 700mm*700mm.

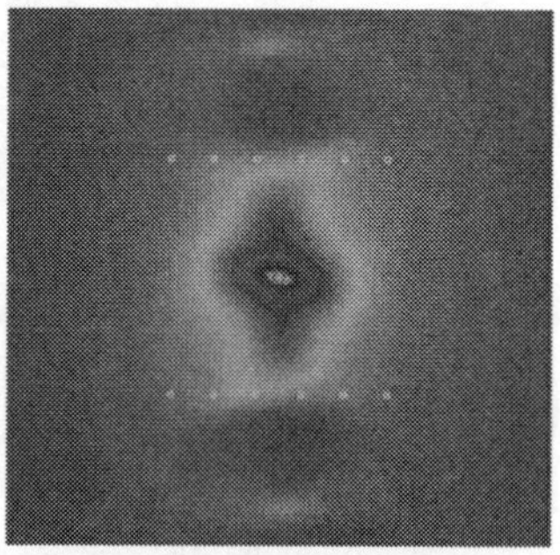

Figure 4c: Hole Ø20mm Plate 1000mm*1000mm.

This reconstruction is based on A0 mode velocity. The consequence of the component size and thus the sensors to edges distance appears clearly: the bigger the distance, the better the defect is located through the imaging and with a higher signal to noise ratio. Whenever it is possible (and depending on the location of the expected defects), it appears then relevant to avoid positioning the sensors too close (with respect to the wavelength) to the specimen edges unless the used imaging algorithm can deal with wave reflections at these edges.

The instrumentation design includes the choice of sensors number and their location on the component. The probability of detection will be reduced if not enough sensors are used, while too many sensors will lead to unacceptable costs and burden for the monitoring industrial deployment. Simulation enables beforehand a study of the performance of potential implementation strategies, quickly and at low cost, without having to instrument real mock-ups, so that only promising solutions can be further tested with real prototypes. For instance, let's compare in Figure 5 the images obtained with 12, 8, 6 or 4 sensors (18mm diameter) when placing them along a circle around the hole in the composite panel. You can observe that the image « quality » remains relevant for 6 sensors and above while it deteriorates significantly for 4 sensors for such layout.

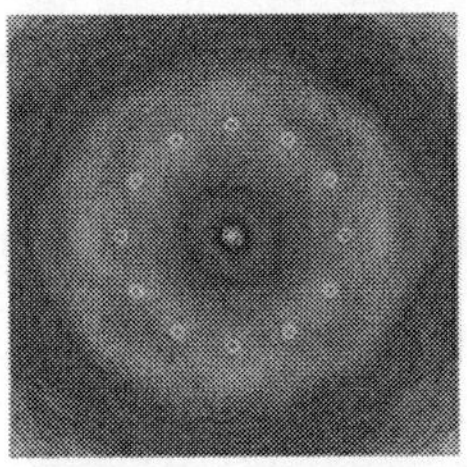
Figure 5a: 12 sensors.

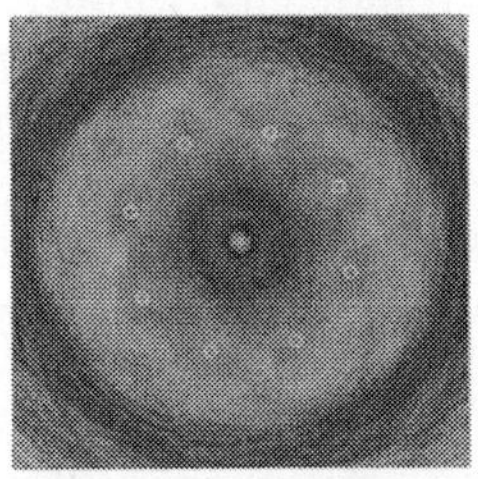
Figure 5b: 8 sensors.

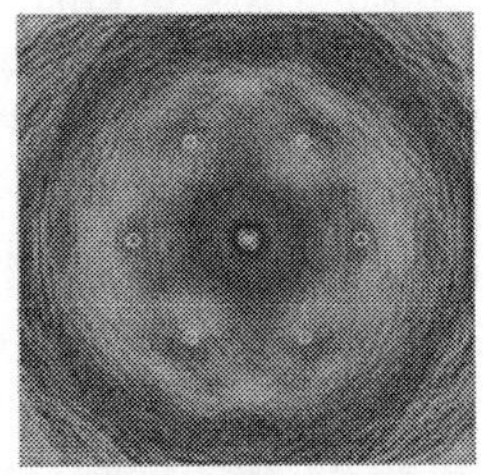
Figure 5c: 6 sensors.

Figure 5d: 4 sensors.

Then, you can observe below in Figure 6 the images obtained with 4 networks of 12 similar sensors but with different layouts: aligned, along one circle, or along staggered circles. You can see that the quality of the defect spot improves (less artefacts so fewer false alarms) when you try to reduce the symmetry of the transducers' arrangement with respect to the component and the targeted flaws.

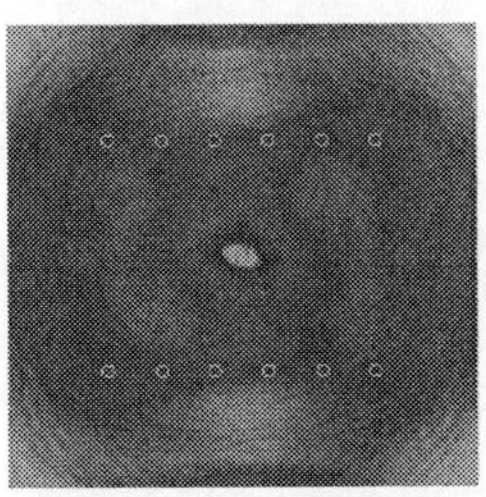
Figure 6a: Linear layout.

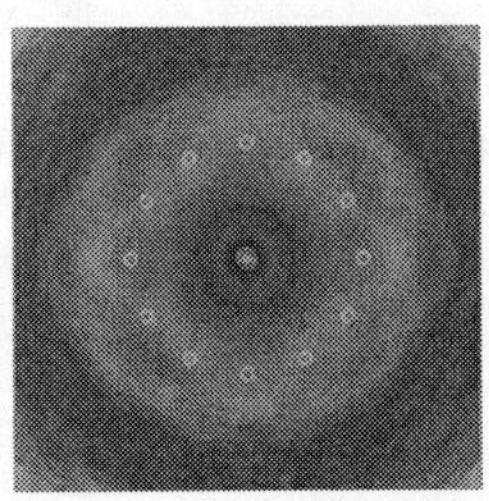
Figure 6b: Circle layout.

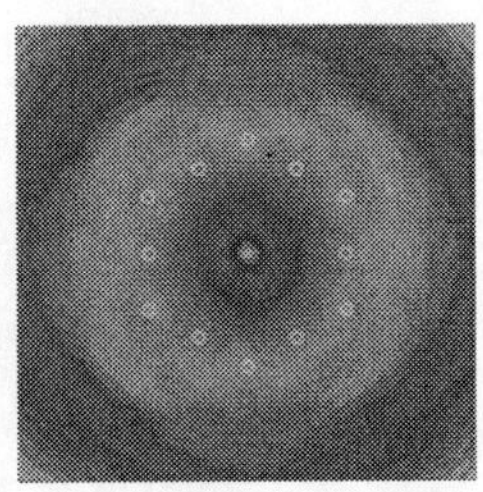
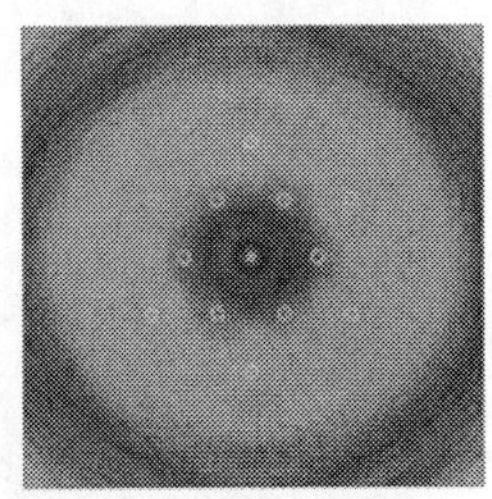
Figure 6c and 6d: 2 staggered circles layout.

Obviously, the monitoring design also includes the choice of the sensors themselves (technology, size, excitation). It is especially fundamental to generate preferentially modes with a wavelength adapted to the defect that you want to detect and maybe characterize. The next simulations focus on the detection of a delamination which is a typical defect in composite multi-layered structures. Such delamination can for instance occur after an impact on the structure during the service life. First, an elliptical delamination of 96mm*48mm and propagated over 6 successive inter-plies has been studied as illustrated below in Figure 7.

Figure 7: Elliptical delamination defined in CIVA SHM.

You can observe below in Figure 8 (the red ellipses represent the true defect size and position) the signatures obtained with this defect type for 2 excitation frequencies (40kHz and 100kHz, respective A0 mode wavelengths of 19mm and 10mm) and 3 sensor diameters (18mm, 10mm, 5mm). At 40kHz, the obtained image gives a clear spot at the defect location but quite badly resolved versus the defect size and shape, independently of the used sensor diameter. With a higher frequency (100kHz), you achieve to obtain an indication closer to the real defect dimensions. And this is further improved for smaller sensors (10 and 5mm diameters) as they preferentially excite the A0 mode at these frequencies, reducing the artefacts from other modes.

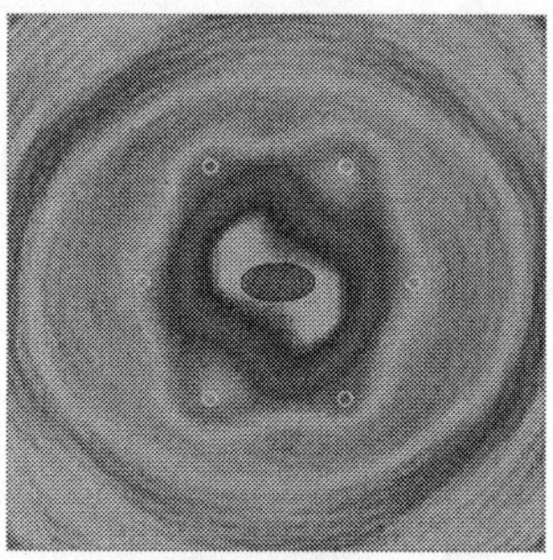

Figure 8a: Frequency 40kHz + Ø18mm.

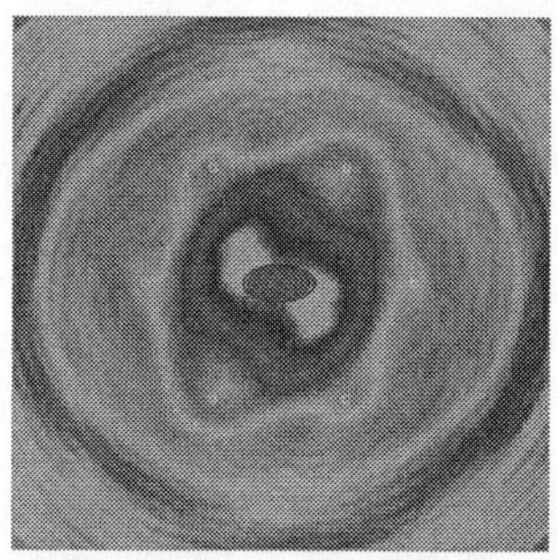

Figure 8b: Frequency 40kHz + Ø10mm.

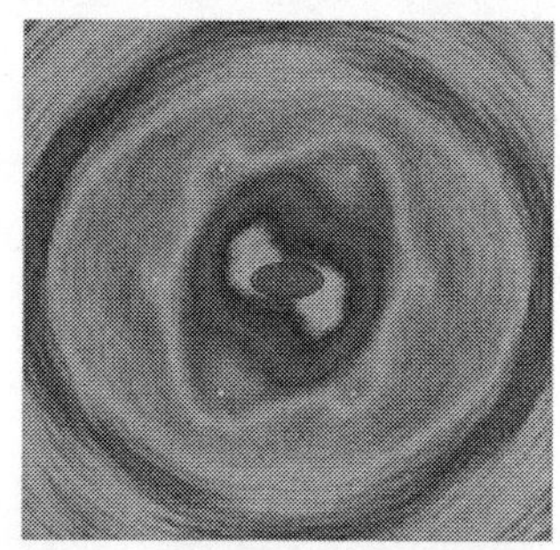

Figure 8c: Frequency 40kHz + Ø5mm.

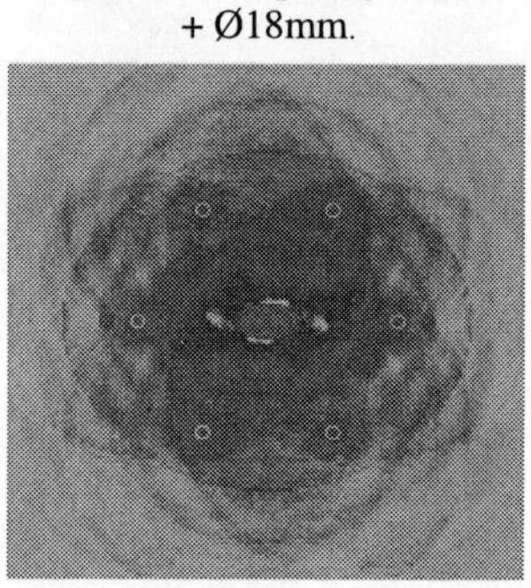

Figure 8d: Frequency 100kHz + Ø18mm.

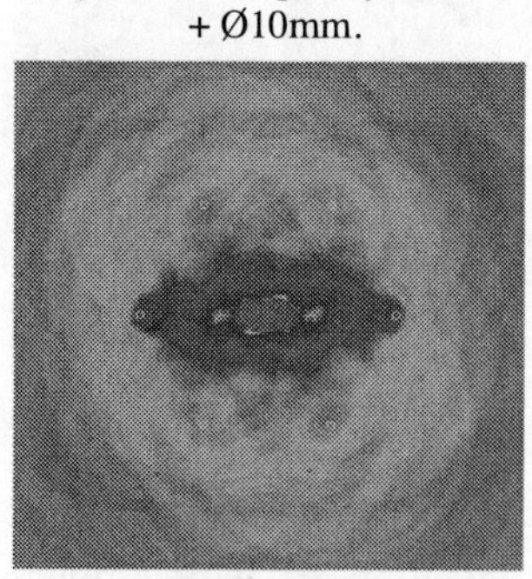

Figure 8e: Frequency 100kHz + Ø10mm.

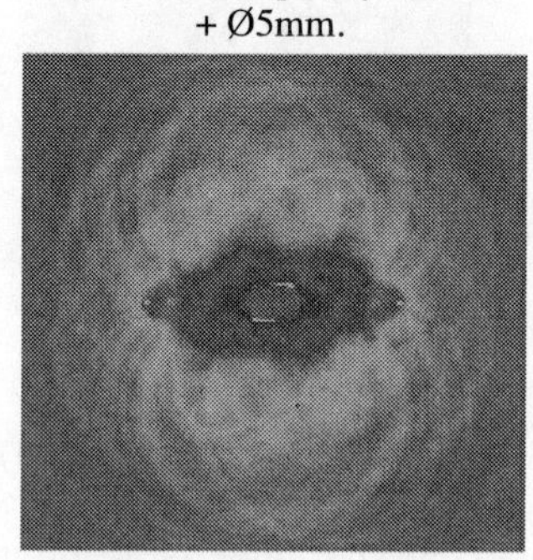

Figure 8f: Frequency 100kHz + Ø5mm.

These simulations prove that a good compromise for this inspection configuration is based on 6 sensors of 10mm diameter and operating at 100kHz to detect and characterize this delamination. Of course, the performances will also strongly depend on the size of the targeted flaw. Below are illustrated the reconstructions obtained with the previous optimal monitoring set up for different elliptical delamination dimensions: 96mm x 48mm, 48mm x 24mm, and 24mm x 12mm.

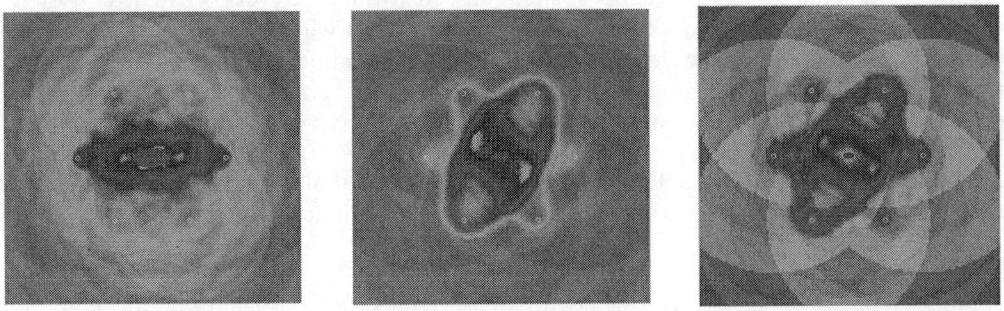

Figure 9a: Delamination 96mm x 48mm.
Figure 9b: Delamination: 48mm x 24mm.
Figure 9c: Delamination 24mm x 12mm.

These simulations show that this setup could remain suitable for smaller delaminations detection (an indication is visible) but the characterization capabilities are noticeably affected as the spot is far less resolved when the defect gets smaller.

The simulation easily provides a better understanding, illustration and justification of performances and limits of a monitoring device for a given targeted defect, from which it is possible to work on optimizations. Here, these images have been obtained assuming the sensors array to be centered around the defect, which is a strong hypothesis. A good knowledge of the materials and the maximum stress zones for a given structure can allow to anticipate preferential areas for defect initiation but it is hard to predict them with a very high accuracy. Once your simulation model has been correctly initiated, it is very easy to modify the defect location and repeat several simulations to study the impact of the flaw position on its detection and characterization with a given SHM setup. It would be much more complex and costly to build as many instrumented mock-ups as possible defect positions with a purely experimental approach.

5. Conclusions

The modelling approach of CIVA SHM as well as its capabilities have been introduced in this paper. After presenting some experimental validation cases, several application examples of CIVA SHM in the context of a damage monitoring in a composite plate have been illustrated. Various parameters have been studied (sensors to specimen edges distance, number and location of sensors, transducer size and frequency, different defect type and dimensions). Simulation represents a powerful tool to help the development and the qualification of monitoring devices. While it does not eliminate the need for experimental trials, prototypes, and mock-ups, it can help reducing the number of needed trials.

More than a competition, a complementarity between simulation and experiment is pointed out: experiments can precisely show the behaviour of real monitoring devices and see the impact of environmental parameters change (structural noise, temperature, aging process, etc.), while simulation can quickly, massively and at low-cost test different monitoring scenarios and predict many damages situations. It would be very costly to design an instrumented mock-up for any defect or sensor set-up cases which would prevent to conduct optimization and performance demonstration studies. Simulation variation studies can help to build large enough data sets when you must describe these performances in a statistical way and work on process qualification.

CIVA SHM provides to the SHM community a dedicated tool which offers very optimized computation times compatible with an intensive usage in an industrial context.

References

[1] https://www.extende.com ; https://www.extende.com/structural-health-monitoring-with-civa

[2] Imperiale, Alexandre, and Edouard Demaldent. "A macro-element strategy based upon spectral finite elements and mortar elements for transient wave propagation modeling. Application to ultrasonic testing of laminate composite materials." International Journal for Numerical Methods in Engineering 119.10 (2019): 964-990.

[3] Mesnil, O., Imperiale, A., Demaldent, E., & Chapuis, B. (2019, May). Validation of spectral finite element simulation tools dedicated to guided wave based structure health monitoring. In AIP Conference Proceedings (Vol. 2102, No. 1, p. 050018). AIP Publishing LLC.

[4] Mesnil, O., Recoquillay, A., Druet, T., Serey, V. Hoang, H. Imperiale, A. & Demaldent, E. (2021). Experimental validation of transient spectral finite element simulation tools dedicated to guided wave based structural health monitoring. Journal of Nondestructive Evaluation, Diagnostics and Prognostics of Engineering Systems, American Society of Mechanical Engineers, In press, 10.1115/1.4050708. hal- 03187419

[5] Moll, Jochen, et al. "Open guided waves: online platform for ultrasonic guided wave measurements." Structural Health Monitoring 18.5-6 (2019): 1903-1914. http://openguidedwaves.de

[6] MICHAELS, Jennifer E. Detection, localization and characterization of damage in plates with an in situ array of spatially distributed ultrasonic sensors. Smart Materials and Structures, 2008, vol. 17, no 3, p. 035035.

Examples of code compliant applications using Total Focusing Method (TFM)

Frederic REVERDY, Gregoire BENOIST, Hubert VOILLAUME, Bertrand BILLEBEAU

Eddyfi Technologies, Les Ulis, France, +33 671427432, freverdy@eddyfi.com

Abstract
Imaging techniques based on ultrasonic Total Focusing Method (TFM) are applied more and more to industrial applications. Adoption of the technique has been made easier with the publication of the revised ASME V and ISO 23864 and 23865 standards. Technique description, equipment standards, personnel qualification and application standards are described in detail allowing operators to prepare their procedures with care. After being the first one to introduce TFM on a portable PAUT system, Eddyfi has improved the performances of the TFM technique in terms of scanning speed and sensitivity while providing code compliant calibration tools. Examples of implementation and verification of FMC/TFM in a procedure according to the ASME and ISO standards will be presented for weld testing and other applications.

Keywords: Ultrasound, Full Matrix Capture (FMC), Total Focusing Method (TFM), Phased-Array (PAUT)

1. Introduction

Phased-array technology has been accepted for many years and used in many NDE applications thanks to its flexibility and the major improvement in productivity. Increasingly, more advanced ultrasonic techniques such as TFM are being implemented. The technique offers superior focusing capabilities and new ways of inspecting components. Its introduction into Standards will help operators develop new procedures. We'll present the principles of TFM, how to optimize a scan plan and all the calibration steps required to perform a TFM inspection.

2. Full Matrix Capture (FMC)/Total Focusing Method (TFM)

The FMC/TFM is an inspection technique that involves two steps. The first one is the data acquisition process called FMC and the second one is the data reconstruction one called TFM. FMC/TFM can be done in post-processing or in real-time, but it is really the later that allowed the democratization of the technique.

2.1 FMC

For this paper, we'll assume that a linear phased-array probe is used. From a Standard standpoint, the FMC is a matrix of n x m signals obtained by firing m times various apertures and receiving m times various apertures for each of these excitations. These apertures both, for emission and reception, can have a different number of elements and use delay laws like focusing or steering. Usually, when referring to FMC, one refers to the elementary FMC for which each transmit and receive pattern is only one element. An elementary FMC obtained from a 64-element probe is composed of 64 x 64 = 4096 signals.

2.2 Plane Wave Imaging (PWI)

PWI is a subset of FMC. It consists in generating a sectorial scan using all the elements at the emission without any focusing, each emission corresponding to one of the n apertures, while reception consists in listening with all the elements individually. For example, if one uses a 64-element probe firing a sectorial angle between 35 and 80° with a step of 5°, the PWI matrix would be composed of 10 x 64 = 640 signals.

Some of the advantages of PWI over elementary FMC are:

- Productivity as the process requires less emissions and has less data to process
- More energy as all the elements of the array are fired simultaneously rather than only one by one
- Less data if one wants to save the PWI matrix for further post-processing

Some disadvantages are:

- The dead zone is increased as the entire aperture of the array is used
- Less omnidirectionality of the energy at emission

Due to the advantages of PWI over elementary FMC for weld inspection, we'll be using PWI for the remainder of this paper.

2.3 TFM

The TFM algorithm consists in coherently summing amplitudes from the n x m signals of the FMC dataset to focus at every pixel of a Region Of Interest (ROI). Mathematically, this can be expressed as:

$$I(P) = \sum_{i,j=1}^{N} s_{ij}\left[t_{ij}(P)\right]$$

where $t_{ij}(P)$ denotes the theoretical time-of-flight corresponding to the propagation time between the emitter E_i and receiver R_j, through one of the pixels P.

The following figure explains the process for one pixel and one signal of the FMC.

1. The TFM process calculates the time-of-flight to go from emitter E_i to pixel P and back to receive R_j. The amplitude corresponding to that time-of-flight is then extracted for that particular signal.
2. The process is then repeated for all the signals of the FMC matrix.
3. All these amplitudes are summed and the result is the amplitude for that pixel in the TFM image
4. The entrire process is then repeated for all the pixels of the image

For a 64-element elementary FMC, the TFM process consists in calculating 64 x 64 = 4096 times-of-flight corresponding to all the possible emission-reception combinations and then extracting the corresponding 4096 amplitudes before summing them to obtain the amplitude of one each pixel. Typical TFM images have dozens of thousand of pixels, we can see how processing intensive the TFM process is.

3. Standards

In 2019, ASME published a revision of Section V "Boiler and Pressure Vessel code" to include the FMC/TFM technique. Definitions and personnel training and experience requirements are specified. Mandatory Appendix XI provides all the requirements for using the FMC/TFM technique: equipment, calibration blocks, calibration tools, examination, evaluation levels, documentation...Nonmandatory Appendix F describes the FMC/TFM technique and how to use it for weld inspection explaining among other things the various modes that can be used, the Amplitude Fidelity calibration...

ISO published two standards 23865 and 23864 associated with the FMC/TFM technique in 2021. ISO 23865 provides guidance for applying ultrasonic testing with arrays using FMC/TFM techniques. This standard provides general definitions, a description of the equipment required, the various calibration steps to respect, comparisons to PAUT technique and examples of the technology applied to industrial applications. ISO 23864 specifies the application of the TFMC/TFM technique to the inspection of welded joints in metallic materials of minimum thickness 3.2 mm.

4. Equipment

4.1 Sample and Calibration mockups

The sample under test here is a 25-mm thick carbon steel V-weld with a 60° bevel and a 30-millimeter (1.2-inch) weld cap. ISO 23864 provides guidance in Annex A on the calibration mockup that can be used.

4.2 Phased-array unit

The PAUT unit used in this paper is a Gekko 64:128 TFM64 from M2M. The unit allows the use of two 64-element probes using FMC/TFM simultaneously. The software is Capture in its 3.3 version.

4.3 Phased-array probes and wedges

We used two 64L5-G3 with S55 wedges. The probes have 64 elements, a pitch of 0.6 mm and a central frequency of 5 MHz. The wedges have a refracted angle of 55° in steel, a length of 73 mm, distance from the front of the wedge to the exist point of 35 mm and a height measured from the center of the array along the focal axis of 34.5 mm

4.4 Scanner

The scanned used is a LYNCS scanner from Silverwing in its weld configuration. The scanner has a resolution of 25.85 points/mm.

5. Scan plan

Codes state that a scan plan with a standard and repeatable methodology for examination should be provided. The scan plan shall include a depiction of the required examination volume coverage, imaging paths, image grid density, weld joint geometry, number of examination scan lines, and search unit placement and movement with respect to the weld axis. For thicknesses greater than 8 mm, the testing volume shall include the weld and parent material for at least 10 mm on each side of the weld.

As mentioned earlier, we'll use PWI method for weld inspection as we are looking for productivity and sensitivity. Capture offers several tools to help operators set their scan plan properly with this technique. The scan plan for PWI, shown in the following figure, allows one to visualize the angles between the incident beam and the weld bevel and to display the near field zone and verify if it is possible to focus the energy.

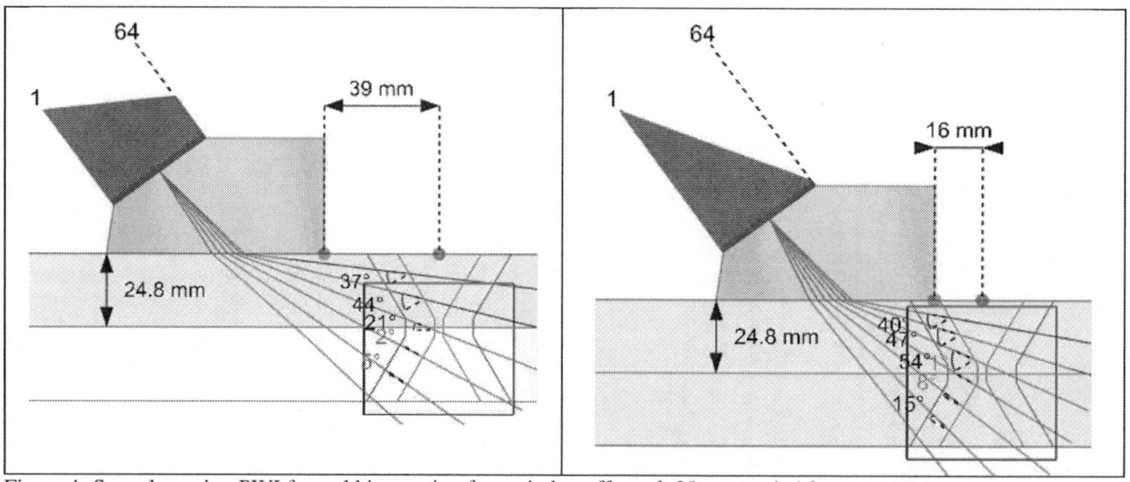

Figure 1. Scanplan using PWI for weld inspection for an index offset of -39 mm and -16 mm

Here are the following parameters that are being optimized to provide maximum detectability, sizing capability and productivity. The exercise is done for one probe, the parameters will be identical for the probe located on the opposite side.

5.1 Index offset

There are two positions possible for TFM for weld inspection. As TFM focuses everywhere, one option is to position the probe as close as possible to the weld cap and benefits from focusing in the first leg to detect tip diffraction echoes from potential defects. Of course, on the second leg the beam is not perpendicular to the bevel anymore. The second position is the same as for PAUT inspection, i.e;, looking to be normal to the bevel on the second leg but we lose the focusing capability on the first leg so have less sensitivity to tip diffraction. For this publication, we'll be looking at both positions -39 mm and -16 mm.

5.2 TFM X offset

The center of the ROI is set along the weld axis at a distance equivalent to the index offset.

5.3 TFM width

The width of the TFM should include the weld, so 30 mm (weld cap) plus 10 mm on each side of the weld as required by standards.

5.4 TFM Z offset

The Z offset is the origin of the ROI along the depth axis, it is at the top of the ROI. Looking at the previous image for an index offset of -39 mm, we can see that the high angles are in the far field, i.e., it is not possible to focus the energy. We could start the Z offset along the top surface but we also want to minimize the number of pixels to optimize the scanning speed. We thus lower the beginning of the ROI where we have focusing along the bevel same side as the probe. For the index offset of -16 mm, the Z offset is 2 mm so we are using a higher ROI.

5.5 TFM height

The TFM height should cover part of the weld on the first leg below the Z offset and the weld on the second leg. We set the height to 44 mm. The top part of the ROI represents the direct mode T-T and the bottom part, i.e., the second leg, represents the indirect mode TT-TT meaning that we are using two TFM modes simultaneously. For the index offset of -16 mm, the height is 52 mm

5.6 Choice of initial and final angles

PWI excitation is a sectorial scan with few angles; it generates plane waves with a width approximately as large as the aperture used. We need to ensure proper insonification of the entire volume of the ROI. We set the initial angle at the left edge of the weld on the second leg (45°) and the final angle close to the top right corner (80°).

5.7 Number of angles

The number of angles is a balance between productivity and sensitivity. If we use a large number of angles, the sensitivity increases significantly as we are sending more energy into the part. However, productivity decreases proportionally with the number of angles. The choice depends on the attenuation of the material and a trial-and-error approach. We found that one step every 7 degrees is a good compromise in this particular case so we are using 6 angles.

5.8 Modes and groups

According to ASME, direct path alone shall not be considered adequate for full volume inspection. We have defined an ROI height that takes into account the first leg for inspection of the root and the second leg for inspection of the bevel and volume of the weld. The first leg is the T-T mode and the second leg is equivalent to the indirect mode TT-TT due to the symmetry with the backwall.

For welded joints, the inspection needs to be carried out from both sides. We use two probes, one on each side of the weld, using a multigroup configuration.

5.9 Pixel size / Number of pixels

The pixel size choice is based on:

- the image resolution: the smaller the pixels, the nicer TFM images look but to the detriment of productivity
- ability to detect the peak amplitude: if pixels are too big, an inspector could simply miss a small reflector, and if the operator uses amplitude-based techniques to size an indication, it may be undersized
- the size of the ROI: the larger the ROI, the more pixels are required, which impacts productivity
- the minimum scanning: encoded inspections can be difficult if the scanning speed is too slow.

To help operators choose the pixel size, Standards have introduced a new calibration called Amplitude Fidelity or Stability. The role of this calibration is to make sure that the pixel size choice is fine enough to guarantee that signal amplitude information is not lost. In our case the pixel size is 0.15 mm.

6. Calibration steps

6.1 Material velocity

The first step in any ultrasonic inspection is to measure the velocity of the material under inspection. For FMC/TFM, this is also an essential step as the time-of-flight calculations describes before use the velocity of the material. We used a conventional S0 V156 at 5 MHz probe to measure the shear velocity of the calibration mockup and the weld sample. We obtained a velocity of 3243 m/s.

6.2 Element check

As we are using PAUT probes, we have to make sure that all the elements are fully functional. According to ISO, the number of dead elements in the active aperture should be less or equal to 1 out of 16 and the dead elements are not adjacent. We are using here a 64-element probes allowing us to have 4 dead elements assuming that they are more than 16 elements apart. We checked our two PAUT probes using the element check tool from Capture and saw no dead element and a sensitivity variation of less than 3 dB.

6.3 Wedge delay calibration

Similarly to the material velocity calibration, the wedge parameters (velocity, angle and height) are essential to extract properly the right times-of-flight from the FMC. An error on one of those parameters would lead to mispositioning of the echoes and variations in amplitude that could lead to missing some defects. We measured the rexolite velocity at 2330 m/s and we performed an angle and height measurement using Capture tool.

6.4 Amplitude Fidelity

When dealing with FMC/TFM, one aspect to consider is the pixel size. It is important not to miss the maximum peak amplitude especially when dealing with methods based on amplitude. The Amplitude Fidelity criterion defines if the FMC/TFM image is properly discretized to preserve amplitude information, i.e. the pixels that compose the FMC/TFM image are small enough to properly measure the signal peak amplitude. ASME V states that amplitude fidelity shall be preserved to 2dB or less. Usually, that criterion can be passed if the pixel size is smaller than λ/6 for RF signals and λ/4.5 for envelope. However, a theoretical approach doesn't take into account many factors such as the real probe central frequency, wedge and material attenuation...ISO 23865 provides an experimental procedure to check if the Amplitude Fidelity is satisfied. This procedure has been implemented in Capture and is used here on the SDH of our calibration mockup to verify if our pixel size selection satisfies the Amplitude Fidelity criterion for three SDH positions. The following figure shows the Amplitude Fidelity tool with the SDH located 2 mm from the left edge of the ROI

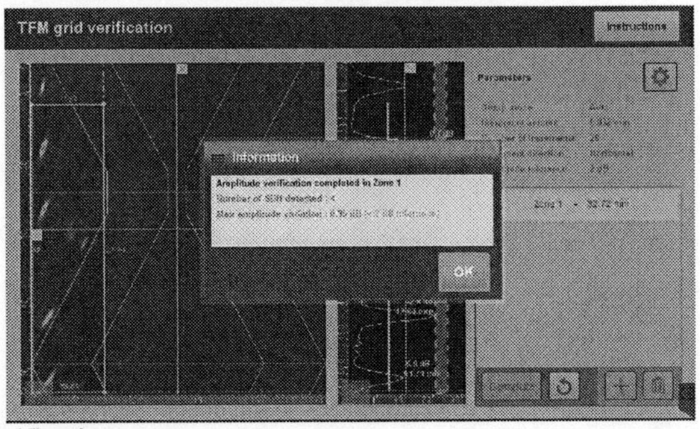

Figure 2. Amplitude Fidelity verification

6.5 Sensitivity correction or Time Corrected Gain (TCG)

For PAUT, standards require that all individual beams used must be calibrated to provide distance measurement and amplitude correction over the sound path employed in the examination. While TFM doesn't exactly create beams, each pixel is the result of a "delay law" applied to an array probe to focus the acoustic energy at that point. Amplitude calibration for TFM requires having the same sensitivity for every pixel in the ROI.

ISO 23865 describes a method on how to perform TCG. Essentially, the probe is moved over side-drilled holes and the amplitude on each SDH is recorded for every horizontal line in the ROI, over its complete width. Then, a gain is applied for every horizontal line and the depth of each SDH to obtain 80% screen height

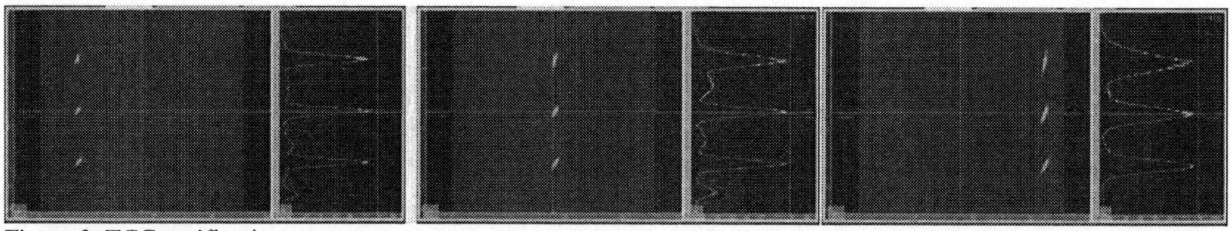

Figure 3. TCG verification

At the end of the TCG calibration, all pixels have the same amplitude, regardless of their position in the ROI. The previous image shows TFM images of three SDH obtained for different mechanical positions. The echodynamic curves project the maxima of each row along the depth axis, allowing one to directly see the peak amplitude of the SDH. We can check that the amplitude of the three SDH is exactly the same for all positions at 80% screen height.

7. Results

Following all the calibration steps, we have acquired data on the welded joint mentioned before using a mechanical step of 1 mm along the scan axis for the two index offset positions of -16 mm and -39 mm. Scanning speed for this multigroup configuration was 100 mm/s.

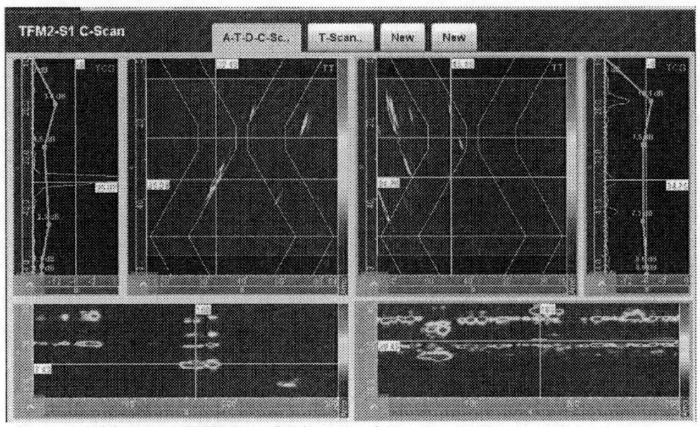

Figure 4. Capture interface for a multigroup TFM weld inspection

Three defects were found in the weld; a lack of sidewall fusion (LOF), a toe crack and a root crack.

LOF:

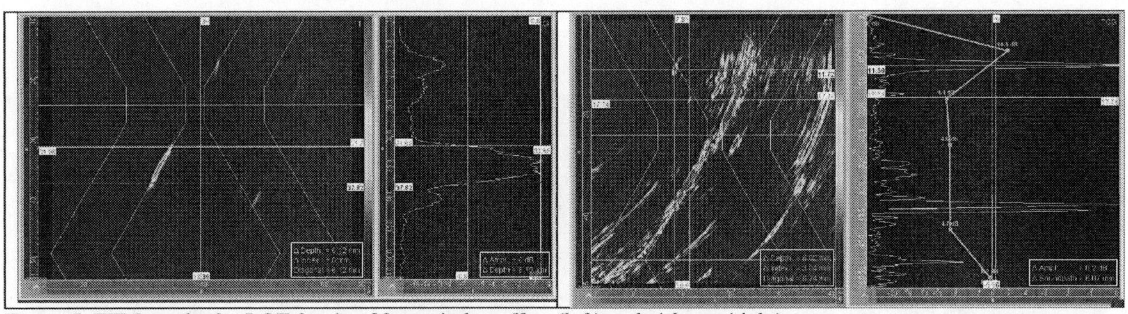

Figure 5. TFM results for LOF for the -39 mm index offset (left) and -16 mm (right)

For the -39mm index offset, sizing is based on a -6 dB decibel drop on the echo on the second leg using the echodynamic along depth and 3 cursors. Measurement is 6.1 mm. Signal-To-Noise (SNR) ratio is 36 dB. For this index offset, we can confirm that this is a planar defect. Due to high focusing with a large aperture (64 elements), this configuration would allow better sizing than PAUT for small defects, minimize oversizing and thus false alarms. The tip diffraction on the first leg have a relatively poor SNR due to the fact that they are not in the near field and cannot probably be used.

For the -16 mm index offset, sizing is based on measuring the maxima of the tip diffraction on the first leg. They don't always appear along the length of the defect, it depends on the orientation of the edges of the defect. When both present, height measurement is relatively easy and precise, in the previous image we measure 6 mm. It was measured at 5.9 mm using a TOFD inspection.

Toe crack:

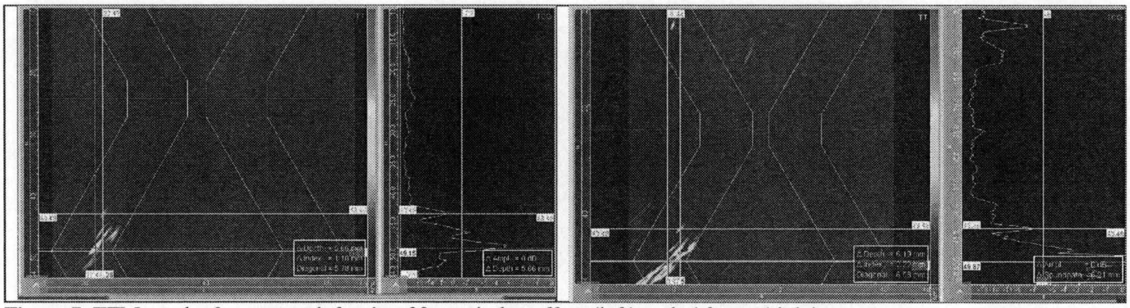

Figure 7. TFM results for toe crack for the -39 mm index offset (left) and -16 mm (right)

For the toe crack we are looking at measuring the distance between the corner echo and tip diffraction on the second leg. Measurement is pretty much the same 5.7 and 6.2 mm for both index offset positions. We can see for the -16 mm index offset that we also obtain a tip diffraction from the bottom of the defect on the first leg. The height of the toe crack was measured at 5.9 mm using TOFD so relatively similar to TFM.

Root crack:

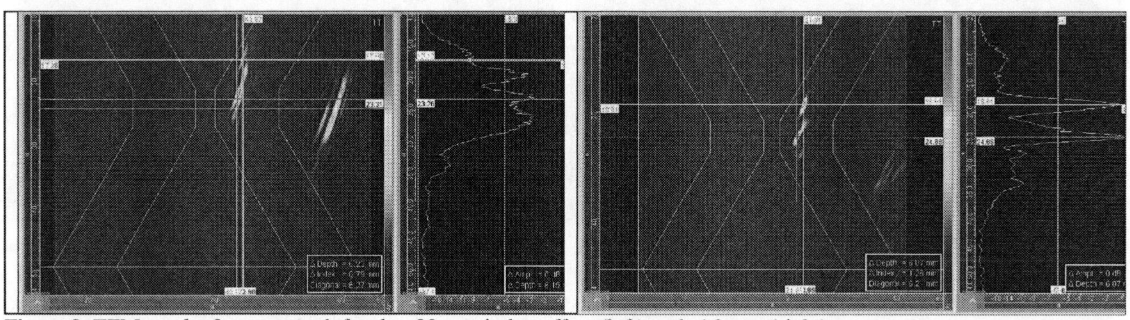

Figure 8. TFM results for root crack for the -39 mm index offset (left) and -16 mm (right)

For the root crack we are looking at measuring the distance between the corner echo and tip diffraction on the first leg. Measurement is pretty much the same 6.2 and 6.1 mm for both index offset positions. It was measured at 6.1 mm with TOFD.

3. Conclusions

Standards are now including TFM as one of the NDT techniques. They provide all the requirements and guidance necessary to develop adequate procedures. We presented in this paper a multigroup weld inspection using TFM following all the calibration steps required by ASME and ISO and at speed (100 mm/s) similar to other ultrasonic NDT techniques used today. Two approaches have been presented, one with an index offset similar to the one used for PAUT and another one close to the weld cap.

Sizing was compared to results obtained using a TOFD configuration. Both gives provide high detectability and sizing capabilities. Due to its high focusing capabilities, TFM allows for better sizing of indications reducing potential need for repair.

Broadband Local Wavenumber Estimation for Damage Detection and Quantification in CFRP Components

Joost SEGERS [1], Saeid HEDAYATRASA [1,2], Gaétan POELMAN [1], Wim VAN PAEPEGEM [1], Mathias KERSEMANS [1*]

[1] Mechanics of Materials and Structures (UGent-MMS), Department of Materials, Textiles and Chemical Engineering (MaTCh), Ghent University, Technologiepark-Zwijnaarde 46, 9052 Zwijnaarde, Belgium

[2] SIM Program M3 DETECT-IV, Technologiepark-Zwijnaarde 48, B-9052 Zwijnaarde, Belgium

*Corresponding author: Mathias.Kersemans@Ugent.be

Abstract
Local wavenumber estimation (LWE) is traditionally applied to a narrowband full wavefield response of the test specimen. The technique proves promising for damage detection in fiber-reinforced polymer (i.e. composite) structures. However, the narrowband nature of the traditional LWE techniques bring several challenges for application on actual test cases.
In this paper, the performance of a novel self-reference broadband version of the LWE technique is investigated. The technique is based on modepass filters in the wavenumber-frequency domain. Compared to the traditional narrowband LWE implementations, the proposed method allows a higher level of automation, removes the need for a priori knowledge on the material and/or defect properties, and results in an improved characterization of defects. The performance is evaluated for an aluminum plate with flat bottom hole defects and for a damaged cross-ply CFRP aircraft panel.
Keywords: Fiber-reinforced polymers, Composites; Non-destructive testing (NDT), Self-reference Broadband Local Wavenumber Estimation, Scanning laser Doppler vibrometry, Guided Waves, Depth quantification

1. Introduction

Composite materials (e.g. carbon fiber-reinforced polymers CFRP) are increasingly used for critical components in several industrial sectors (for example aerospace, automotive). A major challenge is the detection of internal damages in these composites which may have occurred during manufacturing or during operational life. One possibility for damage detection in thin-walled composite structures is to analyze the vibrational response measured on the surface. Many different wave actuation and sensing configurations are possible, combined with advanced data processing methods. In this study, piezoelectric actuators are used for excitation combined with a 3D scanning laser Doppler vibrometer (SLDV) to measure the resulting full wavefield.

In the last decades, multiple methods were developed to convert the full wavefield dataset to a damage map [1-13]. Among others, local wavenumber estimation (LWE) has gained a lot of attention as it seems promising for accurate defect detection and quantification. As the name indicates, the wavenumber of a specific Lamb mode is estimated at each point of the component. For linear isotropic materials, the Rayleigh-Lamb equation (also referred to as the dispersion relation) gives the relation between the local wavenumber k [1/m], the temporal frequency of the wave f [Hz] and the local material thickness d [m] [14]:

$$\frac{\tanh\left(\frac{\beta d}{2}\right)}{\tanh\left(\frac{\alpha d}{2}\right)} = \left[\frac{\left((2\pi k)^2+\beta^2\right)^2}{4\alpha\beta(2\pi k)^2}\right]^s \quad \text{with} \quad \begin{aligned} \alpha^2 &= (2\pi)^2\left(k^2 - \frac{f^2}{c_l^2}\right) \\ \beta^2 &= (2\pi)^2\left(k^2 - \frac{f^2}{c_t^2}\right) \end{aligned} \tag{1}$$

Where c_l and c_t are the material's longitudinal and transverse wave velocity, respectively. The parameter s equals +1 or -1 for the anti-symmetric or symmetric Lamb modes, respectively.
As an example, the dispersion curves of (isotropic) aluminum (c_l = 6420 m/s, c_t = 3040 m/s) are shown in Figure 1. In Figure 1 (a), the solution is given in terms of the wavenumber-thickness product versus the frequency-thickness product, resulting in a single dispersion curve that is valid for any material thickness *d*. In Figure 1 (b), the dispersion curves are shown for an aluminum plate of specified thicknesses. Note that for the A_0 mode, the wavenumber increases when the material thickness becomes smaller. As such, a horizontal crack in the aluminum plate will result in the local increase of the A_0 mode wavenumber as graphically illustrated on Figure 1 (c). In the same way, a delamination defect in a composite plate results in an increased local wavenumber for the A_0 mode.

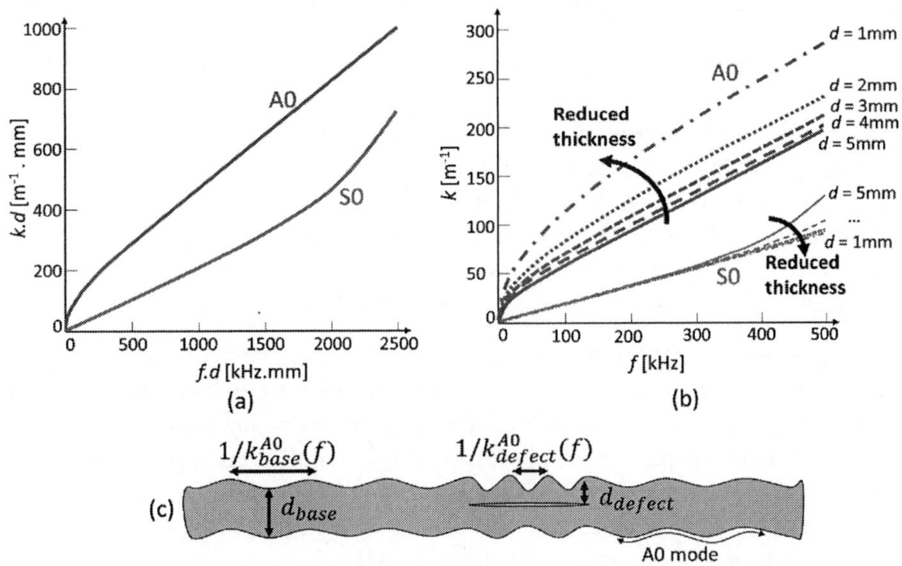

Figure 1: Dispersion characteristics of aluminum material: (a) Dispersion curves expressed as wavenumber-thickness *k.d* in function of frequency-thickness *f.d*, (b) Dispersion curves expressed as wavenumber *k* in function of frequency *f* for several material thicknesses *d* and (c) Schematic illustration of the local change in A_0 mode dispersion behavior at a horizontal crack or delamination.

Multiple LWE algorithms were developed, also resulting in a multitude of other names for LWE such as, acoustic wavenumber spectroscopy, instantaneous wavenumber estimation, multi-frequency LWE, etc [15-19]. All the mentioned LWE approaches have in common that the wavenumber map is obtained corresponding to a specific (center) frequency of excitation and a specific Lamb mode. In most cases, the A_0 mode is chosen because the A_0 mode dispersion curve varies in function of the material's thickness d over the complete frequency axis (see Figure 1 (b)).

Recently, a novel LWE approach is proposed by the current authors which considers broadband excitation and subsequent self-reference mode filtering in the wavenumber-frequency domain [20]. The broadband response of the test specimen is obtained using sine sweep piezoelectric excitation (or pulsed laser excitation), combined with SLDV measurements. The novel self-reference broadband local wavenumber estimation (SRB-LWE) approach shows an enhanced performance on the level of defect evaluation and reduction of the required user input compared to the traditional LWE implementations. A local thickness map is obtained while the only parameter that needs to be known in advance is the base material thickness (or most prominent thickness) of the investigated test specimen. And even in case that also this thickness is unknown, the SRB-LWE will give the relative local thickness compared to the base material.

In this paper, the performance of SRB-LWE is further investigated. First, the measurement procedure and the used test specimens are described. Next, the defect detection and depth identification capability of SRB-LWE is evaluated for a 5 mm thick aluminum plate with flat bottom hole defects and for a 1.1 mm thick CFRP aircraft panel with a disbonded backside stiffener.

2. Materials and Measurements

SRB-LWE is performed on two test specimens (see Figure 2).
The first test specimens is a 400x400x5 mm^3 aluminum plate. Ten flat bottom holes (FBHs) with diameter Ø 25 mm are milled into the backside of the plate. Figure 2 (a) shows the plate together with a table wherein the remaining material thickness d of every FBH is specified. Note that there is one FBH with a remaining thickness of 0 mm that is thus an open hole.

The second component is a CFRP part manufactured for use in the vertical stabilizer of an Airbus A320 aircraft (see Figure 2 (b)). On the backside, stiffeners are visible which are bonded to a base plate. The thickness of the base plate, on which the stiffeners are bonded, measures 1.1 mm. The component is manufactured using a cross-ply layup. The material's elastic properties and densities are unknown. The component was scrapped by the manufacturer because an ultrasonic C-scan inspection revealed a defect at the central stiffener. The time-of-flight (TOF) map obtained from inhouse immersion ultrasonic C-scan inspection using a 5 MHz focused transducer in reflection mode is added to the figure. This C-scan result reveals a relatively large disbond between the central stiffener and the base plate.

Vibrations are introduced through small, low power piezoelectric actuators (Ekulit type EPZ-20MS64W). The actuators are attached to the backside using phenyl salicylate which has the advantage that the actuators can be easily removed by re-melting (> 41 °C) this bonding substance. A broadband sine sweep voltage signal is generated using the SLDV's built-in function generator. The sweep's start and end frequencies are listed in Table 1. The voltage signal is amplified through a Falco WMA-300 high voltage amplifier before it is supplied to the piezoelectric actuator.

The full wavefield velocity response is recorded using a 3D infrared SLDV (Polytec PSV-500 3D Xtra). For each test specimen, the area for which the wavefield is recorded (i.e. scan area) is indicated on Figure 2. All relevant data acquisition settings, i.e. sampling frequency f_s, number of samples, number of averages and scan point spacing are included in Table 1. Only the out-of-plane velocity component $V_Z(x, y, t)$ is used because the SRB-LWE algorithm is operated on the A_0 Lamb mode, which has a dominant out-of-plane surface vibration.

(a)

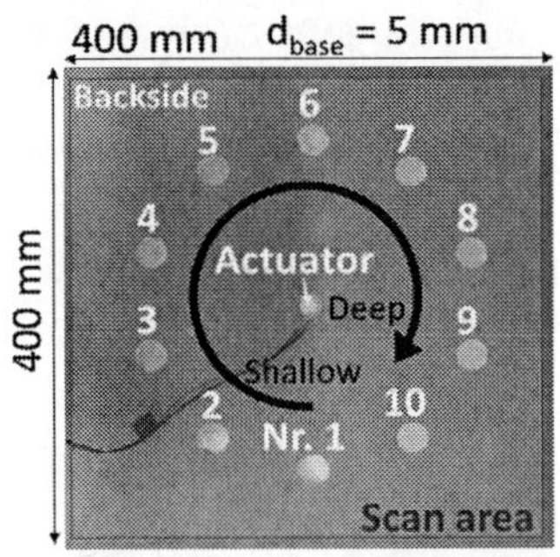

FBH Nr.	1	2	3	4	5	6	7	8	9	10
d (mm)	0	0.3	0.9	1.35	1.9	2.7	2.95	3.3	3.9	4.35

(b)

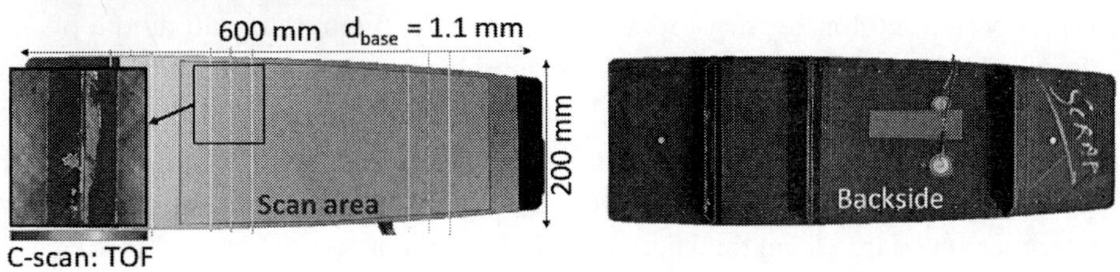

Figure 2: Test specimen: (a) Aluminum plate with 10 flat bottom holes of diameter 25 mm and remaining material thickness *d,* (b) CFRP aircraft panel with disbond at the central stiffener.

Table 1: Characteristics of excitation signals and SLDV data acquisition.

Material	Defect	Excitation			SLDV			
		f_{start} (kHz)	f_{end} (kHz)	V_{pp}	f_s (kS/s)	# samples	# averages	point spacing (mm)
Al 5 mm	FBH Ø 25 mm	5	300	100	1250	10 000	20	3
CFRP 1.1 mm + Stiffeners	Disbond	5	300	50	625	10 000	15	2

3. Self-reference Broadband Local Wavenumber Estimation Algorithm

The SRB-LWE algorithm consists of multiple signal processing steps which are discussed in detail by the current authors in Ref. [20]. A short summary is provided below.

First, the dispersion curves correspond to the damage-free material are identified in the broadband response of the component. The vibrations that are related to modes different from the A_0 mode (i.e. the S_0 mode vibrations) are removed using a modestop filter.
Next, the broadband response is ran through a bank of A_0 modepass filters. Each modepass filter assumes a certain material thickness. The filters are calculated solely based on the identified dispersion curve that corresponds to the damage-free material (i.e. self-reference approach). The bandwidth of the filters is automatically determined.
At last, the local material thickness is identified by looking at the broadband energy (i.e. bandpower) in the modepass filtered velocity fields. The defects are readily identified as local anomalies in the material's thickness.

4. SRB-LWE Results and Discussion

4.1 Aluminum Plate with FBHs

SRB-LWE is performed for the measurement results of the 5 mm thick aluminum plate that contains FBH defects of diameter 25 mm and variable remaining material thickness d (see also Figure 2 (a)). The resulting estimated local thickness map is shown in Figure 3 (a). Figure 3 (b) shows line plots of the estimated local thickness at the location of each FBH defect. The true defect diameter (i.e. 25 mm) and the true remaining thicknesses d are indicated on these graphs using dashed lines. In addition, the estimated and true local thickness at each FBH are listed in Figure 3 (c).

The estimated local thickness map is of high quality and reveals all defects, including shallow defects as well as deep defects with a depth higher than 80% of the base material's thickness. In general, the estimated thickness is in good agreement with the true remaining material thickness as seen in Figure 3 (b) and (c). Note that the thickness is estimated with a resolution of 0.2 mm [20].

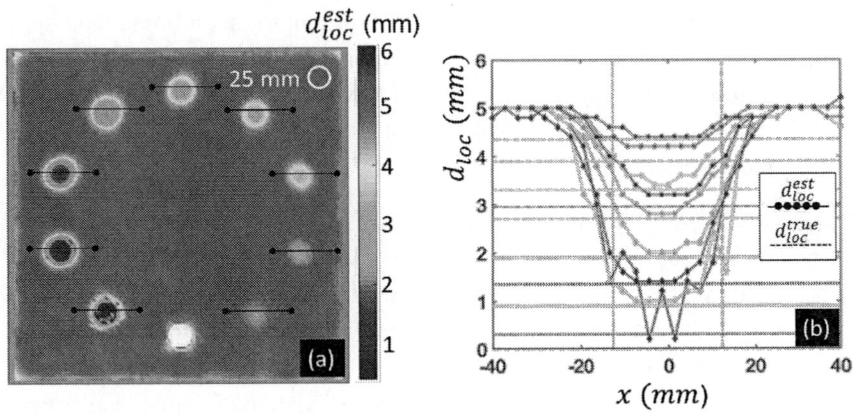

(c)

FBH Nr.	1	2	3	4	5	6	7	8	9	10
d_{loc}^{true} (mm)	\	0.3	0.9	1.35	1.9	2.7	2.95	3.3	3.9	4.35
d_{loc}^{est} (mm)	\	0.2	1	1.4	2	2.8	3.2	3.4	4.2	4.2

Figure 3: Results from SRB-LWE for aluminum plate with FBHs of variable remaining material thickness d, (a) Estimated local thickness map, (b) Estimated local thickness at line through the FBHs and (c) True and estimated local thickness at the FBHs.

4.2 CFRP Aircraft Panel with Disbond

The aircraft panel was manufactured by an industrial partner, and the disbond was unintentionally generated during the manufacturing process. The global wavenumber-frequency map (along $k_y = 0$ m^{-1}) of the measurement is displayed in Figure 4 (a), and the identified dispersion curves of the A_0 and S_0 modes are shown on top with white lines.

Figure 4 (b) shows the TOF map derived from the ultrasonic C-scan inspection. The disbond at the stiffener is revealed by the reduced TOF value. Note that the thickening at the stiffener's middle fin (see also Figure 2 (b)) is not found using the TOF C-scan results.

Figure 4 (c) represents the estimated local thickness map obtained from the proposed SRB-LWE algorithm. For comparison, Figure 4 (d) shows the local wavenumber map obtained from the traditional LWE implementation proposed by Flynn et al. [15]. For this traditional LWE algorithm, a 5-cycle Hanning windowed sine excitation with center frequency 150 kHz was employed, and the algorithm's parameters were selected through trial and error.

The estimated local thickness map obtained through the proposed SRB-LWE algorithm correctly shows the extent of the disbond, i.e. where a local thickness is found similar to the thickness of the base material, i.e. 1.1 mm. A significant improvement in damage identification and evaluation is observed when comparing this local thickness map to the local wavenumber map obtained from traditional LWE. The improved quality is caused by the superior filter efficiency of using a mode filter (instead of a wavenumber filter), by the averaging effect of employing a broadband approach, by the use of a SNR criterion for frequency frame selection and by the self-reference thickness estimation (rather than wavenumber estimation) [20]. Because of this high-quality output of SRB-LWE, it is even possible to estimate the local thickness at the non-defected stiffener's area. The found value of d_{loc}^{est} = 2.20 mm matches perfectly with the true local thickness d_{loc}^{true} = 2.18 mm.

One may notice small discrepancies in the estimated local thickness maps that are caused by the non-isotropy of the composite material [20]. In between the two stiffeners, the true local thickness is 1.1 mm. However, the estimated local thickness map shows a small increase in estimated thickness in the -45° direction around the piezoelectric actuator (see Figure 4 (c)). This local thickness increase must be linked to the layup of the laminate. Therefore, it is expected that the outer plies, which contribute most to the flexural rigidity, have a -45° angle. This expected layup is confirmed after consulting the manufacturer of the aircraft component, who specifies a $[(-45°/+45°)]_{2s}$ layup for the studied component.

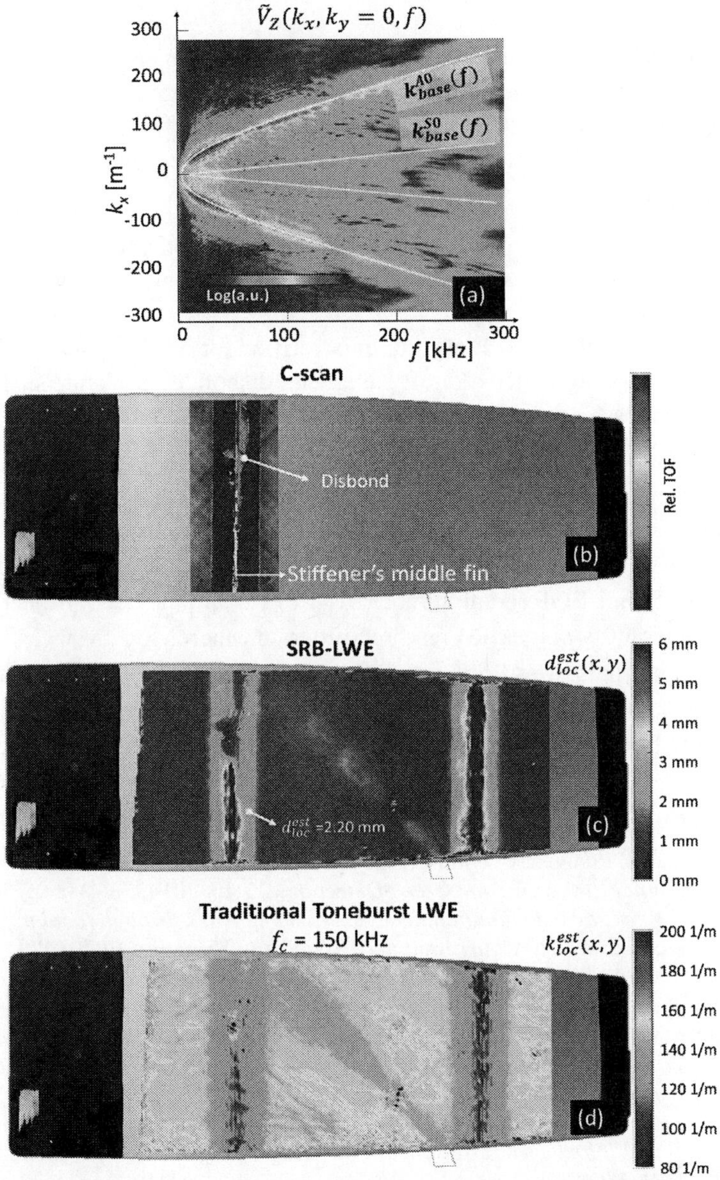

Figure 4: Aircraft CFRP component with disbond at stiffener: (a) Wavenumber-frequency map for the out-of-plane velocity response along $k_y = 0$, together with the identified A_0 and S_0 mode curves, (b) C-scan time-of-flight map at damaged stiffener, (c) SRB-LWE derived thickness map, (d) Traditional LWE [15] derived wavenumber map.

5. Conclusion

The novel full wavefield inspection method: "Self-reference broadband local wavenumber estimation (SRB-LWE)" is used for damage detection in composite components. The procedure results in an estimated local thickness map of the inspected area, in which defects are found as areas of abnormal local thickness. The SRB-LWE algorithm is baseline-free, user-independent and does not require the elastic material properties to be known.

Broadband vibrations are introduced in the component using low-power piezoelectric actuators. A scanning laser Doppler vibrometer records the full wavefield response. The A_0 mode is chosen as the mode of interest for SRB-LWE.

The high performance of the SRB-LWE algorithm is verified for an aluminum plate with flat bottom hole defects, and for a cross-ply CFRP aircraft panel with disbonded backside stiffener. The estimated local thickness maps are found to not only reveal the extent of the defects with high precision, but to also give accurate depth estimation of the defects.

Acknowledgements

The authors acknowledge Fonds voor Wetenschappelijk Onderzoek FWO (grant number 1148018N), Bijzonder Onderzoeks Fonds BOF (grant number 01N01719) and the SBO project DETECT-IV (Grant no. 160455), which fits in the SIM research program MacroModelMat (M3) coordinated by Siemens (Siemens Digital Industries Software, Belgium) and funded by SIM (Strategic Initiative Materials in Flanders) and VLAIO (Flemish government agency Flanders Innovation & Entrepreneurship). The authors further thank SABCA Limburg for providing the CFRP aircraft panels.

References

1. Ostachowicz, W., M. Radzieński, and P. Kudela, *50th Anniversary Article: Comparison Studies of Full Wavefield Signal Processing for Crack Detection.* Strain, 2014. **50**(4): p. 275-291.
2. Segers, J., et al., *Backside delamination detection in composites through local defect resonance induced nonlinear source behaviour.* Journal of Sound and Vibration, 2020. **479**.
3. Segers, J., et al., *Nonlinear Elastic Wave Energy Imaging for the Detection and Localization of In-Sight and Out-of-Sight Defects in Composites.* Applied Sciences, 2020. **10**(3924).
4. Segers, J., et al., *Robust and baseline-free full-field defect detection in complex composite parts through weighted broadband energy mapping of mode-removed guided waves.* Mechanical Systems and Signal Processing, 2021. **151**.
5. Radzienski, M., et al., *Damage Identification in Various Types of Composite Plates Using Guided Waves Excited by a Piezoelectric Transducer and Measured by a Laser Vibrometer.* Sensors (Basel, Switzerland), 2019. **19**(9).
6. Keshmiri Esfandabadi, Y., et al., *Full Wavefield Analysis and Damage Imaging Through Compressive Sensing in Lamb Wave Inspections.* IEEE transactions on ultrasonics, ferroelectrics, and frequency control, 2018. **65**(2): p. 269-280.
7. Cuomo, S., et al., *Machine learning for impact detection on composite structures.* Materials Today: Proceedings, 2020.

8. Song, H. and Y. Yang, *Super-resolution visualization of subwavelength defects via deep learning-enhanced ultrasonic beamforming: A proof-of-principle study.* NDT & E International, 2020. **116**.
9. Spytek, J., et al., *Multi-resolution non-contact damage detection in complex-shaped composite laminates using ultrasound.* NDT & E International, 2020. **116**.
10. Sha, G., et al., *Guided wavefield curvature imaging of invisible damage in composite structures.* Mechanical Systems and Signal Processing, 2021. **150**.
11. Wu, Z., S.Y. Chong, and M.D. Todd, *Laser ultrasonic imaging of wavefield spatial gradients for damage detection.* Structural Health Monitoring, 2020.
12. Kudela, P., M. Radzieński, and W. Ostachowicz, *Identification of cracks in thin-walled structures by means of wavenumber filtering.* Mechanical Systems and Signal Processing, 2015. **50-51**: p. 456-466.
13. Kudela, P., M. Radzienski, and W. Ostachowicz, *Impact induced damage assessment by means of Lamb wave image processing.* Mechanical Systems and Signal Processing, 2018. **102**: p. 23-36.
14. Rose, J.L., *Ultrasonic Guided Waves in Solid Media.* 2014: Cambridge University Press.
15. Flynn, E.B., et al., *Structural imaging through local wavenumber estimation of guided waves.* NDT & E International, 2013. **59**: p. 1-10.
16. Rogge, M.D. and C.A. Leckey, *Characterization of impact damage in composite laminates using guided wavefield imaging and local wavenumber domain analysis.* Ultrasonics, 2013. **53**(7): p. 1217-26.
17. Jeon, J.Y., et al., *2D-wavelet wavenumber filtering for structural damage detection using full steady-state wavefield laser scanning.* NDT & E International, 2020. **116**.
18. Jeon, J.Y., et al., *Damage detection on composite structures with standing wave excitation and wavenumber analysis.* Advanced Composite Materials, 2017. **26**(sup1): p. 53-65.
19. Mesnil, O., C.A.C. Leckey, and M. Ruzzene, *Instantaneous and local wavenumber estimations for damage quantification in composites.* Structural Health Monitoring, 2014. **14**(3): p. 193-204.
20. Segers, J., et al., *Self-Reference Broadband Local Wavenumber Estimation (SRB-LWE) for defect assessment in composites* Mechanical Systems and Signal Processing, 2021. **163**.

Use of Non-destructive Techniques to Detect Damage Induced by Corrosion of Pre-stressing Steel Strands

Josef STRYK[1], Monika MANYCHOVÁ[2], Marta KOŘENSKÁ[1], Aleš FRÝBORT[1], Václav SVOBODA[3]

[1]CDV -Transport Research Centre, Líšeňská 33a, 636 00 Brno, Czech Republic, +420 549 429 330; josef.stryk@cdv.cz, korenska.m@email.cz, ales.frybort@cdv.cz
[2]Brno University of Technology, Faculty of Civil Engineering, Veveří 331/95, 602 00 Brno, Czech Republic; manychova.m@fce.vutbr.cz
[3]Preditest Ltd, Pod višňovkou 23, 140 00 Praha 4, Czech Republic; svoboda@preditest.cz

Abstract

The aim of this work is to provide confirmation of the possible use of two non-destructive techniques to detect damage of pre-stressing steel strands due to corrosion. For this purpose accelerated corrosion procedure has been conducted on these strands in laboratory conditions. Corrosion consequences were analyzed for various duration of accelerated corrosion process. The ultrasonic techniques and metal magnetic memory (MMM) method were used in this experimental study. The results show the application potential of ultrasonic techniques to analyze the effect of damage induced by accelerated corrosion on the nonlinear elastic behaviour of a steel strands. A new parameter, based on sensing of the magnetic field strength in the form of residual magnetization on the surface of the material, has been used in this study. Appraising parameters of measurement by the metal magnetic memory method were presented in the form of magnetogram – dependence of the magnetic field strength and field gradient on the measured distance. The changes of the magnetic field strength proved different stress concentration zones in the steel strands due to corrosion consequences. Subsequently, the differently corroded strands were subjected to tensile tests in order to determine their mechanical properties to verify and complement results of non-destructive tests.

Keywords: Pre-stressing steel strands, electrochemical corrosion, cracking, nonlinear ultrasonic testing, metal magnetic memory method.

1. Introduction

Pre-stressed concrete is a complex material taking advantage of the combination of concrete and steel reinforcement. There is a need to study how pre-stressed concrete structures react to damage processes in the build environment. Steel reinforcement corrosion belongs to these damages very often. This paper is a continuation of our previous work on the analysis of damage effects induced by corrosion on the linear and nonlinear elastic properties of a pre-stressing steel strand during the accelerated corrosion process [1]. Having gained first results and experience we switched our focus on combination of two non-destructive methods for this purpose. The ultrasonic techniques and metal magnetic memory (MMM) method were used in our experimental study. The aim of this contribution is to analyze the effects damage induced by corrosion on the nonlinear elastic properties and metal magnetic memory parameters of the prestresing steel strand after accelerated corrosion process. Precisely the combination of these two methods may contribute to a great deal to further improvement of the defectoscopy and analysing steel corrosion consequences in reinforced concrete structures.

2. Preparation of specimens for experiment

Commercially available seven-wire steel strand of 15.7 mm in diameter was object of our experiments. The strand consists of a steel central straight wire surrounded by six helical wires. Eight strand pieces of 1.3 m in long were the subject of the experiment. The strands were divided into two groups, labelled L9 and L10. One piece was left in its original state and three pieces of strands were subjected to different degrees of electrochemical corrosion in each group. The duration of the corrosion process was set for individual strands for 2, 6 and 12 hours. The strands labels were following: L90, L100 - strands were not subjected to corrosion, L92, L96, L912 and L102, l06, L1012 - strands according to the duration of their accelerated corrosion. The electrochemical corrosion process has been described in detail in [1] and is briefly described here. The cylindrical corrosion cell was replaced by a box with a removable top, allowing visual inspection of the progressive corrosion of the strand during the corrosion process, which was accelerated by an electric current of a constant value of 3 A. The strand central part of 50 mm length was only subjected to corrosion; remaining steel surface in corrosion box has been coated with an electric insulator. An illustration of the corrosion process realization, including the detail of the corrosion box is shown in Fig. 1.

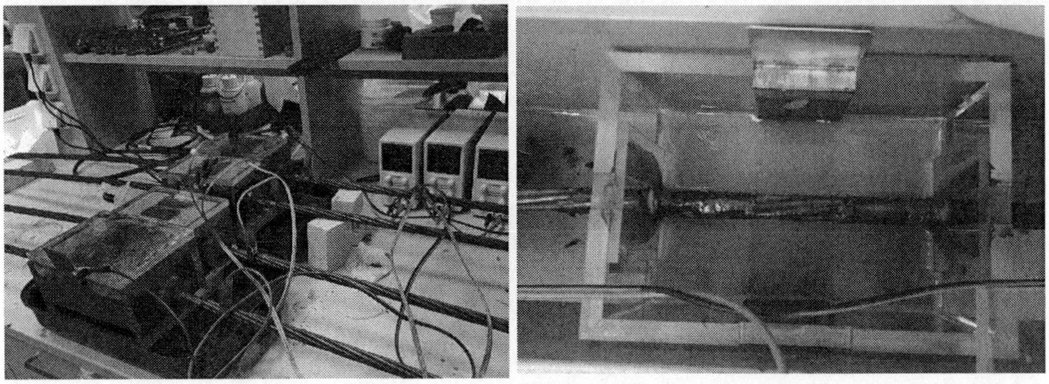

Figure 1. Illustrative record of prestressing strands subjected to accelerated corrosion: on the left - four strands exposed to corrosion process; on the right - detail of the corrosion box

The corroded strands were measured by non-linear ultrasonic spectroscopy in CDV laboratories and after that had drive transport to Preditest company for MMM measurement and tensile testing. The visual condition of all measured strands is documented in Fig. 2. The strands are sorted according to the increasing duration of the corrosion process: L90 to L912; L100 to L1012.

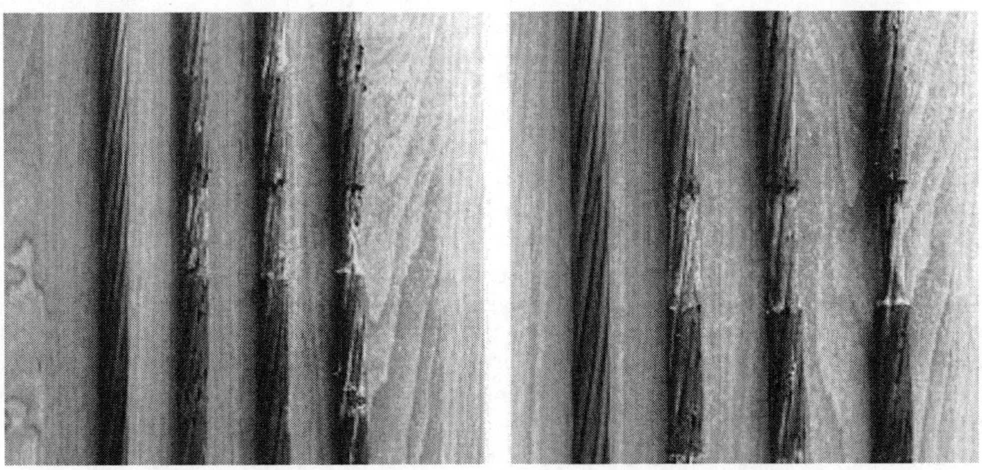

Figure 2. Two groups of measured strands: on the left L9 group; on the right L10 group

2.1 Nonlinear measurements

As results, when an ultrasonic wave propagates through material and interacts with the micro-structural defects, non-linear effects linked to these higher order waves are generated. As the micro-damage grows, non-linearity should increase. Nonlinear dynamic response may manifest itself in a variety of manners.

Nonlinear methods focusing on the material response nonlinearity have been described in many papers for example [2]. A single harmonic ultrasonic signal measurement method which has been applied to the experiment is briefly described below.

This method analyses the effect of nonlinearities on acoustic signals propagating through the specimen. The nonlinearity gives rise to additional signals featuring different frequencies according to Fourier expansion. In general, the amplitudes of these additional components decrease with natural number n. In this case, where the excitation is accomplished by a single frequency f_1 (Fig. 3) the nonlinearity gives rise to other harmonic signals, whose frequencies f_v obey the Fourier series formulas:

$$f_n = n \cdot f_1 \quad \rightarrow \quad \Big| \, n = 0, 1, 2, \ldots \infty \qquad (1)$$

In general, these frequency component amplitudes are falling when the harmonic order natural number, n, is increasing. If the nonlinearity effect is not entirely symmetrical, there can arise low-amplitude second and higher even-numbered harmonic components, whose amplitudes may be lower than those of the odd-numbered ones. Among these emerging components, the third harmonic is the most distinctive one, see Fig. 3. Therefore, its amplitude is pursued most frequently [3].

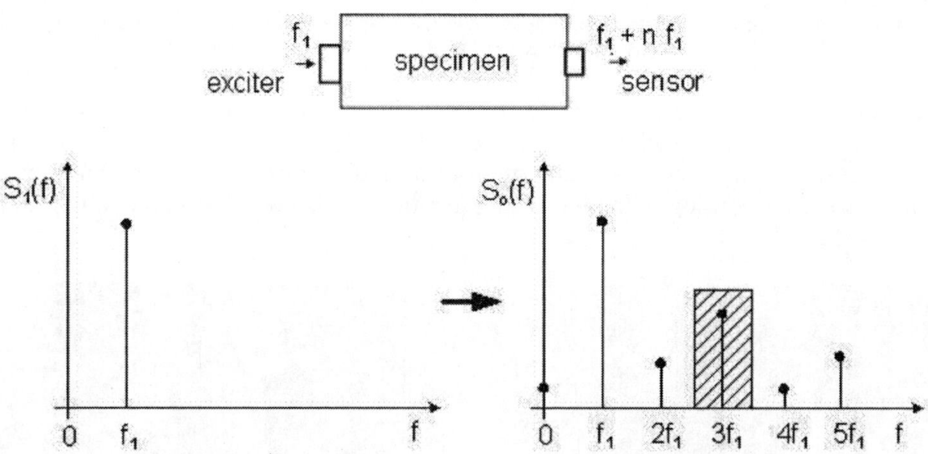

Figure 3. Frequency spectrum of a nonlinear medium response

In the framework concerned the degradation of the strands, in the act of the corrosion process consequences, is linked to a nonlinear mechanical behaviour. As a result, when an ultrasonic wave propagates through the strand and interacts with microstructural defects, nonlinear terms linked to these higher order waves are generated. As the microdamage grows, nonlinearity should increase.

Two different strategies have been used in this study to put evidence the appearance of nonlinear features due to damage of distinct corroded strands. First, the observation of the frequency spectrum of the received signal, which may allow direct observation of the higher harmonic generation or intermodulation distortion features, see Fig. 3. In second place the nonlinearity parameter can be derived from the amplitudes of the fundamental and harmonic frequencies. Assuming that changes in the wave propagation velocity and the attenuation are small, and in experimental conditions like those used in this work (nonlinear measurements done with an input signal of fixed frequency, and the transducers located always at the same positions), the parameter of nonlinearity can be approximated as: $Bpn = A_3 / A_1^3$, being A_1 amplitudes of the fundamental and A_3 third harmonic waves [4]. In practice, the variation of these ratios relative to the initial undamaged conditions is used as an indicator of the microstructural damage [5].

2.1.2 Measurement results

Measuring results from the first strategy are represented by the following Fig. 4 showing the values of higher harmonic frequency amplitudes as a percentage of the 1st harmonic (exciting frequency f_1) amplitude for both groups.

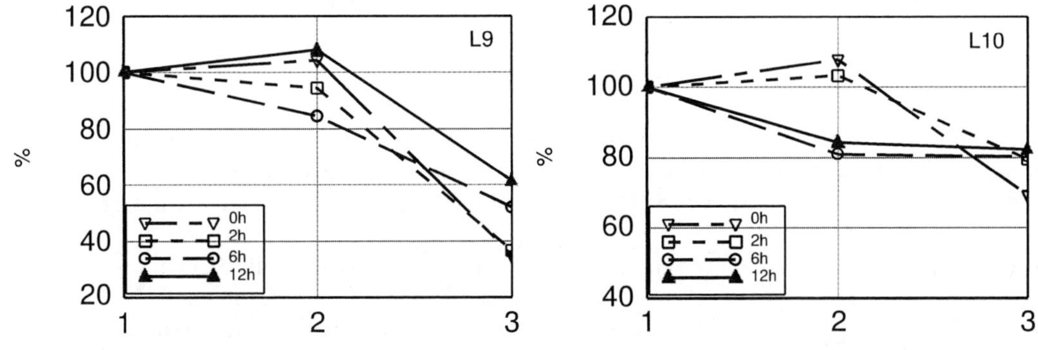

Figure 4. Higher harmonic frequency amplitudes expressed as a percentage of the first harmonic amplitude; on the left - group L9, on the right - group L10

It is apparent from Fig. 4 that the amplitudes of the third harmonics (A_3) are always lower than the fundamental amplitude (A_1) in all frequency spectra. Anomalies are evident in some spectra; the second harmonic amplitude (A_2) exceeded the fundamental amplitude in the case of non-corroded strands L90 and L100, strand L912 (after 12 hours of corrosion process action) and strands L102 (after 2 hours of corrosion process action). The consequences of corrosion manifested by a growing in the amplitude values of the third harmonics after 6 and 12 hours of corrosion action, as can be seen in both graphs in Fig. 4. This observation point out to the possibility that nonlinear ultrasonic measurements may provide an early warning of the damage due to strand corrosion.

Figure 5 shows the variations of the relative values of the parameter *Bpn* from tests of strands which were subjected corrosion process for differed time (0, 2, 6 and 12 hours). Any value in the figure is normalized to its corresponding value at zero time (0 hours). A linear increase in the value of this parameter is evident from 2 hours of the corrosion process. It must be recalled that the values of this *Bpn* parameter is proportional to those of the rigorously defined nonlinearity parameters, Sect. 2.1. Previously it was proposed a tenfold increase relative to the initial values of these parameters, as the arbitrary threshold for considering a critical increase of the nonlinear elastic features, which in turn would be indicative of the presence of significant defects or damage in the material medium [6].

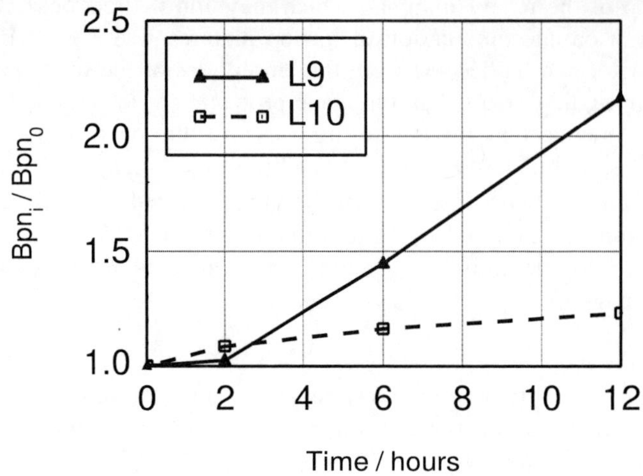

Figure 5. Evolution of the parameter *Bpn* normalized to this initial values (at the time = 0 hours)

2.2 Metal magnetic memory method

The metal magnetic memory (MMM) method is an NDT method based on sensing of the magnetic field strength *Hp* in the form of residual magnetization on the surface of the material. It uses one or more probes, each sensing the magnetic field in three on perpendicular planes, see Fig. 6. The residual magnetization of the material is affected by the production process, machining, cutting, welding, bending, heat treatment, cooling, operational stress, corrosion of the material, etc. A very important factor influencing the magnetic memory of a material is the degradation process caused by the operational stress and conditions to which the structure is exposed.

In the case of steel reinforcement in a concrete beam, the degradation processes are further approached by corrosion processes, stress corrosion and possible negative effects from the assembly process, which in turn reduce the life of the object. The MMM method use can detect these effects in the form of a magnetogram, i.e. the dependence of the magnetic field strength on the distance of the probe *Lx* from the beginning of the measurement.

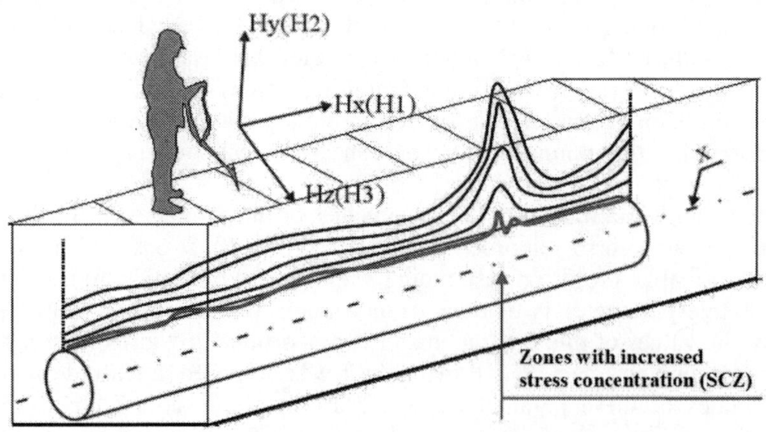

Figure 6. Scheme of measurement by the metal magnetic memory method above a steel pipe in the ground in the longitudinal direction [7]

The MMM method can detect these effects from a magnetogram, i.e., dependence of the magnetic field strength on the distance of the probe from the beginning of the measurement.
A special cart, on which the probes are placed, is moved above the surface of the measured area (e.g., in a pipe or beam axis) and the wheel of the device measures the driven distance. The signal from the each individual channels, together with the distance, is recorded in digital form in the device's memory and displayed graphically either directly as *Hp* or in the form of gradient *dHp/dx.* The recorded data are presented in the form of a graph (so-called magnetogram). Based on the evaluation, we can then determine the zones with increased stress concentration (SCZ), where there is an increased probability of changes or defects in the material structure. The stress concentration is proportional to the measured magnitude of the magnetic field strength gradient around the given position.
A significant advantage of this NDT method is the measurement speed and also the high sensitivity. Used measurement technique has been described in [7].

2.2.1 Measurement results

The steel strands were arranged in the line on the ground, Fig. 7. The indicative contactless measurement was applied in the distance 100 mm from the surface of the strands.

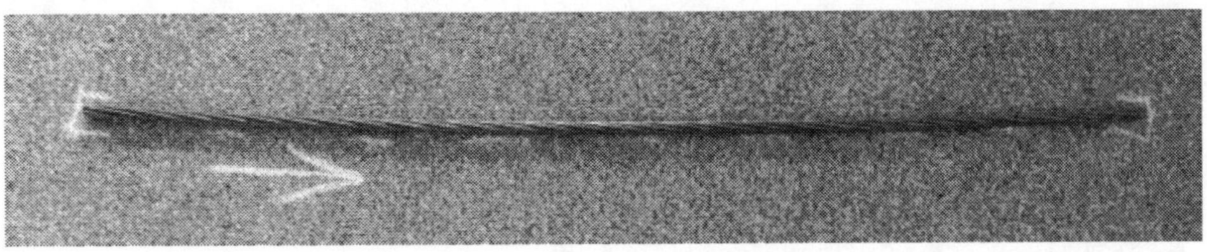

Figure 7. Intact steel strand L90

The measurement was repeated three times to verify the repeatability of the measurement results. The measurement results are presented here in the form of the magnetograms from measurements of group L9. Black rectangle indicated in magnetograms corresponds to evaluated strand central part subjected to corrosion process. A magnetogram in Fig. 8 illustrates the results of the continuous scanning of intact (uncorroded) strand L90.

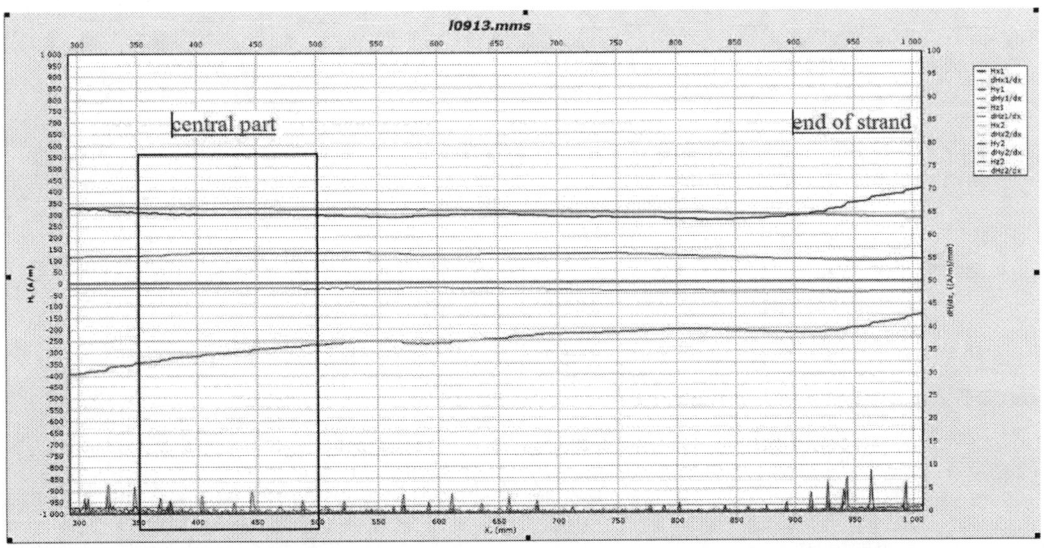

Figure 8. Magnetogram 10913.mms corresponds to undamaged steel strand L90

The magnetogram shows graphically the field strength *Hy* and the field gradient *dHy/dx* as a function of the distance *Lx*. The uniform distribution of the field strength *Hy* in the magnetogram confirms that the structure of the strand is uniform throughout its length.
Fig. 9 presents the magnetogram from scanning the strand L92 after 2 hours of corrosion process. The magnetogram shows zone of slight increased stress concentration of field strength in the range of coordinates x, correspond to corroded part of strand L92. Max value of the field gradient *dHy/dx* amounts to 9.0 (A/m)/mm.

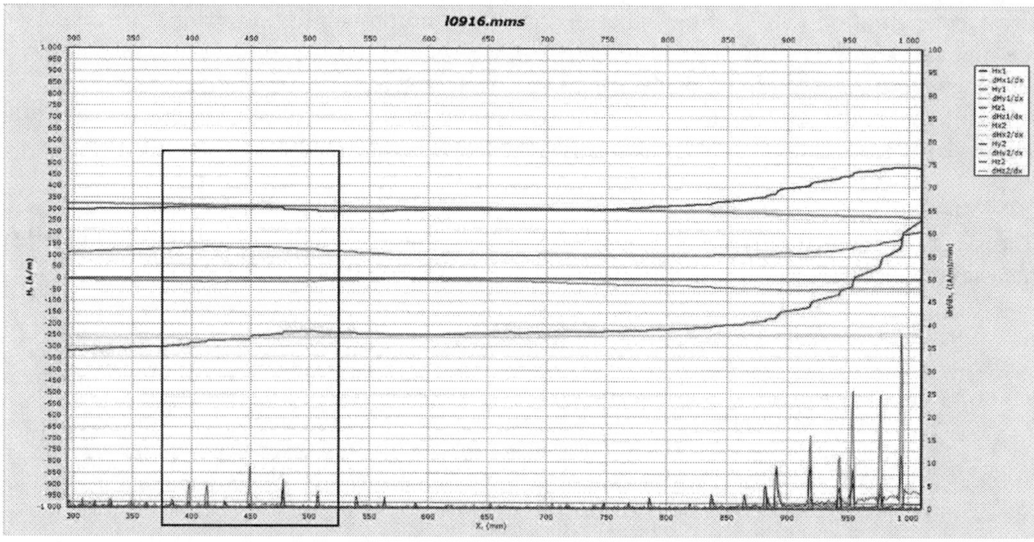

Figure 9. Magnetogram 10916.mms correspond to strand L92 – after 2 hours of corrosion process

Next Fig. 10 presents the results of the continuous scanning of strand L96, after 6 hours of corrosion process. In the magnetogram is evident further increased stress concentration of field strength in comparison with magnetogram in Fig. 9. Max value of the field gradient *dHy/dx* amounts to 43.0 (A/m)/mm.

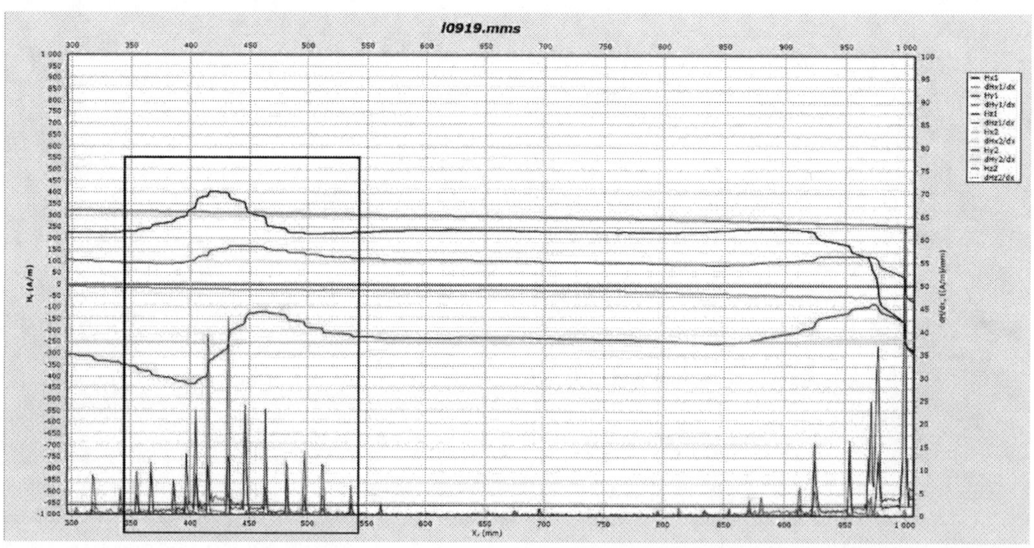

Figure 10. Magnetogram 109169.mms correspond to strand L96 – after 6 hours of corrosion process

Fig 11 presents measurement results of strand L912 after 12 hours of corrosion process. The magnetogram shows higher value of the field gradient *dHy/dx* in comparison with strand L96. It amounts to 100.0 (A/m)/mm.

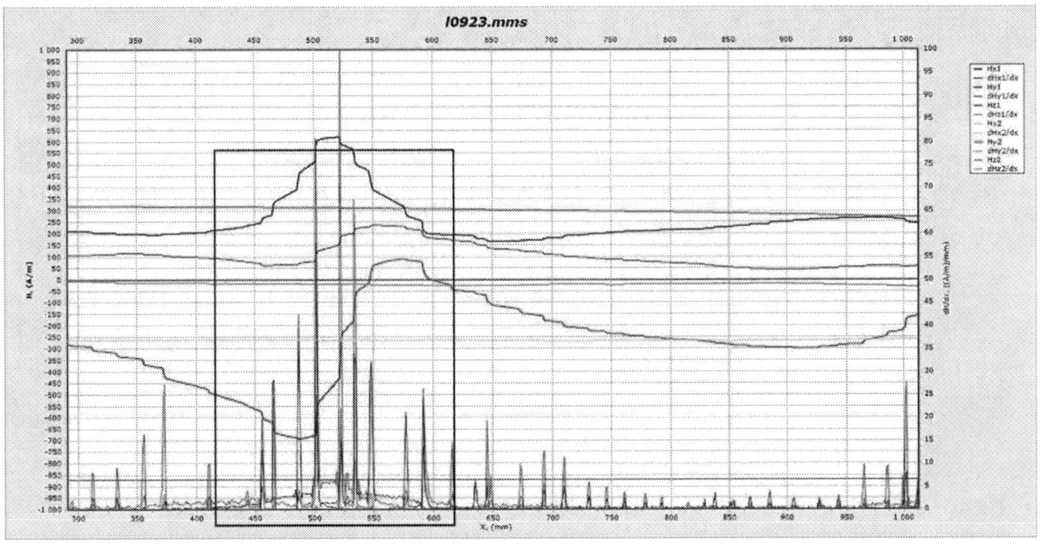

Figure 11. Magnetogram 10923.mms correspond to strand L912 – after 12 hours of corrosion process

The values of the field gradient *dHy/dx* are graphically shown for both groups (L9, L10) in Fig. 12.

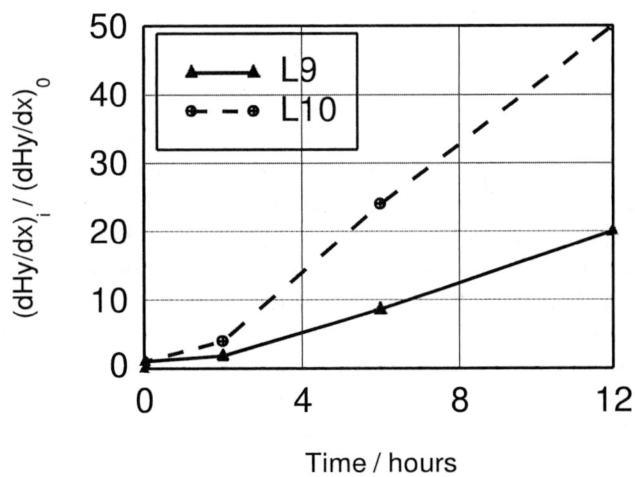

Figure 12. Evolution of the parameter *dHy/dx* normalized to this initial values (at the time = 0 hours)

Graph shows the variations of the relative values of the parameter *dHy/dx* from tests of strands which were subjected corrosion process at differed time (0, 2, 6 and 12 hours). Values in the figure are normalized to its corresponding value at the time = 0 hours. A linear increase in the value of this parameter is evident from 2 hours of the corrosion process action, similarly as in the case of parameter *Bpn* in Fig. 5.

Finally the tested strands were subjected to tensile strength tests in order to determine their mechanical properties and verify and complement measurement results. Appraising parameters of strand damage were following: normalized values of strand diameters in affected area and force F_{max} corresponding to failure of strands during the tests. These parameters are compared graphically in Fig. 13.

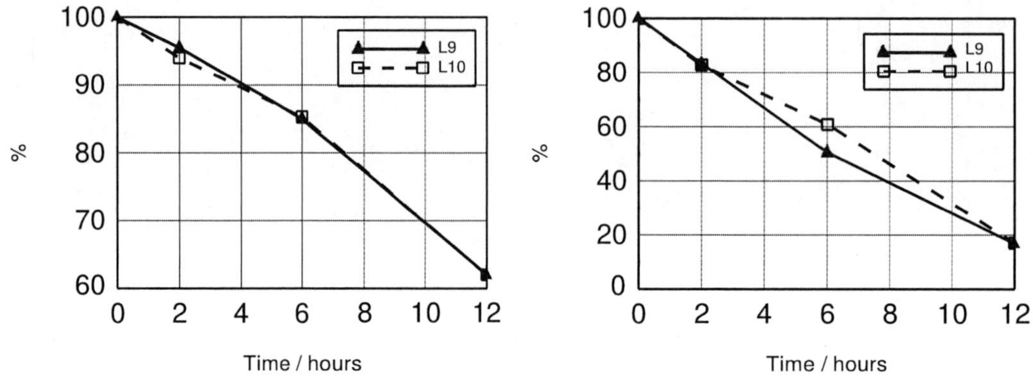

Figure 13. Consequences of corrosion on strands: on the left - normalized values of strand diameters; on the right - normalized values of force F_{max} corresponding to failure of strands during tensile strength tests

Both graphs show a deterioration of mechanical properties of corroded strands. The normalized values of the parameters (strand diameter and breaking force F_{max}) show a linear decrease with increasing time of subjecting the steel strands to the corrosion process.

3. Conclusions

The results obtained in this work provide confirmation that it is possible to use ultrasonic techniques and metal magnetic memory method for the damage detection of pre-stressing strands caused by corrosion. In cause of ultrasonic techniques damage consequences were observed in the form of strong nonlinear features of the received signal. Harmonic distortion was clearly observed. The parameter of nonlinearity was approximated as $Bpn = A_3 / A_1^3$. The variation of these ratios relative to the initial undamaged condition was used as indicator of the structural damage. Regarding the results corresponding to both specimen groups (L9, L10) the relative parameter values increase with time of corrosion process. The growing trend is linear from two hours duration to twelve hours of accelerated corrosion process.

Similar trend is evident in the case of use of magnetic memory method and observation of the magnetic field gradient *dHy/dx.* Its magnitude shows similarly linear increase from two hours duration until twelve hours of the corrosion process action.

Finally, the strands were subjected to tensile strength tests in order to determine changes of their mechanical properties due to impaired diameter of strands as a consequence of corrosion. The values of strand diameters and corresponding forces F_{max} at failure of the strands proved linear decrease with increasing corrosion.

More research is necessary to confirm the findings and interpretations of this study, and to advance on the proposal of practical procedures for routine engineering surveys of pre-stressed or post-tensioned concrete structures suspected to be affected due to steel corrosion.

Acknowledgments

This paper was prepared as a part of research project No. CK01000108, which was financed from the state budget by the Technology agency of the Czech Republic and the Ministry of Transport of Czech Republic under the DOPRAVA 2020+ Programme.

References

1. J. Stryk, K. Pospisil, M. Korenska, A. Frybort, M. Manychova, 'An Initial Investigation on the Potential Applicability of Ultrasonic Testing to Assess Effects of Corrosion on Linear and Nonlinear Elastic Behaviour of Pre-stressing Steel Strands', In Proceedings of Xth NDT in Progress, October 2019, pp. 75-84.
2. M. Korenska, M. Manychova, L. Pazdera, 'Experimental Study of the Nonlinear Effects Generated in a Concrete Structure with Damaged Integrity', Russian Journal of Non-destructive Testing, 2013, 49(9), pp. 530-537.
3. K. Hajek, J. Sikula, 'Testing of Low-Current Contacts Quality and Reliability by Using Third Harmonic Distortion', In Proceedings of 45th- IEEE Holm Conference, pp. 211-213, September 2003.
4. D. R. Raichel, 'The Science and Applications of Acoustics', AIP Series in Modern Acoustics and Signal Processing, 2006, pp. 625-626.
5. A. A. Shah & Y. Ribakov, 'Non-linear ultrasonic evaluation of damage concrete based on higher order harmonic generation', Materials & Design 30, 2009b, pp. 4095-4102.
6. M. A. Climent, M. Miró, J. Carbajo, et al., 'Use of non-linear ultrasonic techniques to detect cracks due to steel corrosion in reinforced concrete structures', *Materials* 2019, *12*(5), 813; https://doi.org/10.3390/ma12050813.
7. K. Pospisil, M. Manychova, J. Stryk, et al.: 'Diagnostics of reinforcement condition in concrete structures by GPR, metal magnetic memory method and impact-echo method', Remote Sensing, 2021, 13, 952. pp. 1-15; http://doi.org/10.3390/rs 13050952.

AUTHOR INDEX